AF847658

Springer Series in Surface Sciences

Volume 71

Series Editors

Gerhard Ertl, Fritz-Haber-Institut der Max-Planck-Gesellschaft, Berlin, Germany

Hans Lüth, Peter Grünberg Institute, Forschungszentrum Jülich GmbH, Jülich, Germany

Roberto Car, Department of Chemistry, Princeton University, Princeton, NJ, USA

Mario Agostino Rocca, Dipartimento di Fisica, Università degli Studi di Genova, Genova, Italy

Hans-Joachim Freund, Fritz-Haber-Institut der Max-Planck-Gesellschaft, Berlin, Germany

Shuji Hasegawa, Department of Physics, University of Tokyo, Bunkyo-ku, Tokyo, Japan

This series covers the whole spectrum of surface sciences, including structure and dynamics of clean and adsorbate-covered surfaces, thin films, basic surface effects, analytical methods and also the physics and chemistry of interfaces. Written by leading researchers in the field, the books are intended primarily for researchers in academia and industry and for graduate students.

More information about this series at http://www.springer.com/series/409

Hideaki Kasai · Allan Abraham B. Padama ·
Bhume Chantaramolee · Ryan L. Arevalo

Hydrogen and Hydrogen-Containing Molecules on Metal Surfaces

Towards the Realization of Sustainable Hydrogen Economy

Springer

Hideaki Kasai
Department of Precision Science
and Applied Physics
Graduate School of Engineering
Osaka University
Suita, Osaka, Japan

Bhume Chantaramolee
National Institute of Technology
Akashi College
Akashi, Hyogo, Japan

Allan Abraham B. Padama
Institute of Mathematical Sciences
and Physics
College of Arts and Sciences
University of the Philippines Los Baños
Los Baños, Laguna, Philippines

Ryan L. Arevalo
Department of Physics
University of San Carlos—Talamban
Campus
Cebu City, Philippines

Balik Scientist Program
Department of Science and Technology
Taguig, Philippines

ISSN 0931-5195 ISSN 2198-4743 (electronic)
Springer Series in Surface Sciences
ISBN 978-981-15-6993-7 ISBN 978-981-15-6994-4 (eBook)
https://doi.org/10.1007/978-981-15-6994-4

© Springer Nature Singapore Pte Ltd. 2020
This work is subject to copyright. All rights are solely and exclusively licensed by the Publisher, whether the whole or part of the material is concerned, specifically the rights of translation, reprinting, reuse of illustrations, recitation, broadcasting, reproduction on microfilms or in any other physical way, and transmission or information storage and retrieval, electronic adaptation, computer software, or by similar or dissimilar methodology now known or hereafter developed.
The use of general descriptive names, registered names, trademarks, service marks, etc. in this publication does not imply, even in the absence of a specific statement, that such names are exempt from the relevant protective laws and regulations and therefore free for general use.
The publisher, the authors and the editors are safe to assume that the advice and information in this book are believed to be true and accurate at the date of publication. Neither the publisher nor the authors or the editors give a warranty, expressed or implied, with respect to the material contained herein or for any errors or omissions that may have been made. The publisher remains neutral with regard to jurisdictional claims in published maps and institutional affiliations.

This Springer imprint is published by the registered company Springer Nature Singapore Pte Ltd.
The registered company address is: 152 Beach Road, #21-01/04 Gateway East, Singapore 189721, Singapore

To our families,
loved ones, and friends.

Preface

To unravel and alter the physical properties of materials at the atomic level are among the themes of our long years of research works. We do quantum dynamical calculations to realize the 'intelligent design' of materials. Intelligent design develops relevant and novel materials aimed for a specific purpose. The mechanism is described by the so-called computational materials design (CMD). In CMD, mathematical models are used to simulate the system of interest, and the obtained data are analyzed to establish physical mechanisms. Knowing how a system works through these physical mechanisms is essential in materials design and functionalization. It can be used to alter certain conditions and parameters to generate a novel or modified system. Experimental groups could then verify the findings of the computational investigations. By adopting this process, several groundbreaking results have been developed, e.g., in the fields of renewable energy and environmental applications.

Hydrogen-based technology is one of the many areas we explore in the field of intelligent design research. This topic encompasses many areas, which includes hydrogen storage and transport. The use of light elements such as hydrogen for fuel requires a particular type of handling. We have worked on hydrogen storage-related topics, such as ortho-hydrogen (o–H_2) to para-hydrogen (p–H_2) conversion. At room temperature, there is a 3:1 ratio for o–H_2 and p–H_2. However, at thermal equilibrium conditions, in the liquid phase of hydrogen, p–H_2 is more stable and dominant. Conversion of o–H_2 to p–H_2 entails giving off heat; hence, there is a possibility of boil-off. In our research, we have taken into account H_2 quantum dynamics by performing density functional theory-based calculations and perturbative calculations to analyze the enhancement of o–H_2 to p–H_2 conversion via surface interaction. The method we proposed is essential to avoid H_2 boil-off problem that arises and works by filling storage tanks with an equilibrium composition of 100% p–H_2.

Apart from this, we are also engaged in the development of electrode and electrolyte materials for fuel cells, particularly proton-exchange membrane fuel cells (PEMFCs). We have worked on finding alternatives to the expensive Pt-based electrode catalyst. We tapped non-precious metal catalysts like nature-inspired hydrogenase for the hydrogen oxidation reaction (HOR) at the anode side and porphyrin- and pyrrole-based materials for the oxygen reduction reaction (ORR) at the cathode side. This will greatly reduce the cost of practical implementations of PEMFCs and will contribute significantly to the development of renewable energy sources. Another important investigation related to catalysis on surfaces is on the quantum tunneling effects of O_2 dissociative adsorption on metal surfaces. The physical principles underlying the relatively low operating temperature of PEMFCs (80–100 °C) have not been understood in past theoretical research, the 'energetics' of the involved processes in particular. However, we propose the importance of the dynamics, in which reactions are analyzed by the motion of the nuclei based on quantum mechanics, and not only on electrons. Consequently, it was found that the quantum tunneling effects of O_2 play an important role for dissociative adsorption on solid surfaces, e.g., Pt, which is now widely used as a cathode catalyst for PEMFC. With this, the necessity of taking quantum effects into account for the analysis of corresponding reactions is being realized. For that purpose, we developed our own computation codes, NANIWA series, NANIWA Dynamics, and NANIWA Statics. These works are just a few of those dedicated efforts to solve the mysteries of materials science with the purpose of contributing significantly to the preservation of the environment through developing novel materials for the realization of renewable energy.

The vision to utilize hydrogen as energy carrier is now a reality. The fulfilment of this vision is a result of years of intensive investigations and collaboration between different sectors of the society. On December 11, 2019, Kawasaki Heavy Industries, Ltd. launched the liquid "SUISO Frontier" hydrogen carrier/ship at Kobe Works. The company will extract hydrogen from brown coal that will be mined at Victoria, Australia. They will liquefy the extracted hydrogen, load it in a carrier, and transport it to Japan. In particular, "the thermos" that contains liquid hydrogen is made at the Harima Works. To maintain the stable liquid state, it is necessary to control the nuclear spin of hydrogen. Such research endeavor is the topic of a joint research and development between researchers of Kawasaki Heavy Industries, Ltd. and the research group of Dr. Hideaki Kasai. Dr. Yamashita, and Dr. Yajima of Kawasaki Heavy Industries, Ltd., and Dr. Kasai of Akashi College, National Institute of Technology participated in the launching ceremony.

The success of fundamental research works, as demonstrated above, relies on one's dedication to understand the properties of and processes happening in materials. This book is written to provide knowledge on the basic concepts and research findings on the interaction of hydrogen and hydrogen-containing molecules with surfaces. The book starts with discussions of the basic ideas of hydrogen reactions on surfaces. An overview of density functional theory is introduced at the start of Chap. 1 to provide the readers with robust theoretical background of first-principles calculations. It is followed by technical discussions of crystal structures, chemisorption, and physisorption processes, and ends with scattering reactions on surfaces. In Chap. 2, the discussion is focused on the behavior of hydrogen and hydrogen-containing molecules on metal surfaces. This includes the concepts of water and methane bond activation, and surface segregation. The quantum effects in hydrogen dynamics on surfaces were discussed in Chap. 3, with emphasis on the formalism and recent developments in this field. Finally, the book ends with a chapter on the review of the current status of hydrogen economy, which provides a scientific, social, and economic perspective of hydrogen economy and its realization in our modern time. In this chapter, the current technologies for hydrogen conversion, storage, and delivery are reviewed, in the context of their implementation following the hydrogen roadmaps of different countries around the world. As such, this book bridges activities of the science community and the implementation of its findings by the different sectors of the society.

"SUISO Frontier" (the liquid hydrogen carrier)

Suita, Japan — Hideaki Kasai
Los Baños, Philippines — Allan Abraham B. Padama
Akashi, Japan — Bhume Chantaramolee
Cebu City, Philippines — Ryan L. Arevalo

Acknowledgements

H. Kasai would like to thank the students and staff for the wonderful years they have shared with him.

B. Chantaramolee acknowledges the Ministry of Education, Culture, Sports, Science and Technology (MEXT) of Japan, Quantum Engineering Design Course (QEDC) of Osaka University, and Tanaka Kikinzoku Kogyo (TKK). He also would like to acknowledge Dr. Mamoru Sakaue and Dr. Elvis Flaviano Arguelles for their valuable support.

A. A. B. Padama acknowledges the Institute of Mathematical Sciences and Physics, College of Arts and Sciences, University of the Philippines Los Baños; the research grants of the University of the Philippines System (ECWRG 2017-2-011; EIDR-C08-003; BPhD-2015-07); the Japan Student Services Organization (JASSO) Follow-up Research Fellowship; the National Institute of Technology, Akashi College; and the Osaka University.

R. L. Arevalo acknowledges the Balik Scientist Program of the Department of Science and Technology, through the Philippine Council for Industry, Energy and Emerging Technology Research and Development.

The authors thank Koizumi Shinichi and Taeko Sato for their continued support and help in realizing this book. The authors also thank the staffs of Springer-Nature Tokyo Office who are involved in the publication of this book.

Contents

About the Authors

Prof. Hideaki Kasai is currently serving as the president of the National Institute of Technology, Akashi College in Hyogo, Japan. He is also a professor emeritus at Osaka University in Osaka, Japan. In his long and multi-awarded career, he has published 483 peer-reviewed works in reputable scientific journals, 13 co-authored books, and been granted 32 patents. Among his most notable contributions to science are his pioneering works on ultra-fast quantum processes, elementary processes in excitations and reactions at surfaces and interfaces, quantum first-principles-based design of nano-material devices, hydrogen storage and fuel cell catalysts, nanospintronics, and properties of strongly correlated systems. He currently holds several key positions in the government and scientific organizations. He is currently a scientific adviser to the Supreme Court of Japan; advisory board member of the Institute of Solid State Physics Supercomputer; board member of The Vacuum Society of Japan, member of the editorial board and deputy editor of the Journal of Physics: Condensed Matter by the Institute of Physics; council adviser of the Physical Society of Japan; board adviser of the Physics Department De La Salle University, academic adviser for the development of Bandung Institute of Technology, and member of the editorial board of the Computational Materials Science.

Dr. Allan Abraham B. Padama is Professor at the Institute of Mathematical Sciences and Physics, University of Philippines Los Banos, Philippines. He completed his BS Physics for Teachers degree at Philippine Normal University—Manila in 2004. He earned his Master of Engineering and Ph.D. Engineering degrees at the Department of Applied Physics, Osaka University, Japan, in 2011 and 2014, respectively. His research activities and scientific publications are focused on computational investigations of gas–surface reactions and of different processes occurring on surfaces facilitated by the presence of adsorbates. He is also interested in designing materials for energy- and environment-related applications.

Dr. Bhume Chantaramolee is currently a specially appointed assistant professor at the National Institute of Technology, Akashi College, Japan. He received his Ph.D. from the Quantum Engineering Design Course (QEDC) of Osaka University under the supervision of Prof. Hideaki Kasai and Prof. Wilson Agerico Diño. He was a sandwich program student at the National Institute of Technology, Akashi College, under the supervision of Prof. Hideaki Kasai and Prof. Hiroshi Nakanishi. His research focuses on the first-principles approach to investigate reactions on catalysts' surfaces.

Dr. Ryan L. Arevalo is currently a Balik Scientist Program grantee of the Department of Science and Technology in the Philippines, and formerly a specially appointed assistant professor at the National Institute of Technology, Akashi College, Japan. He received his Ph.D. in quantum engineering design at Osaka University under the supervision of Prof. Hideaki Kasai. His research has focused on first-principles calculations for heterogenous catalysis.

Chapter 1
Reactions on Surfaces

Abstract The reaction of atoms or molecules on surfaces is the cornerstone of surface science and is of profound importance in many physical, chemical, and biological phenomena. Understanding the basic mechanism of these surface reactions lay the foundation of sophisticated ideas that give rise to diverse industrial and technological applications that are important in our modern time. This chapter will discuss the mechanism of reactions between atoms or molecules on surfaces from the viewpoint of theory and simulation. The chapter starts with a brief discussion of density functional theory, followed by a review of basic crystallography, and ends with the theory of adsorption, desorption, and scattering. These basic concepts of surface science are indispensable in a more complex and thorough analysis of more complicated systems that will be discussed in the succeeding chapters.

1.1 Density Functional Theory

First principles calculations based on density functional theory (DFT) are widely used to investigate the fundamental properties of surfaces and atomic-scale mechanism of surface reactions. Such method has successfully clarified the bond-breaking and bond-forming mechanisms of industrially relevant reactions such as hydrogen dissociation, oxygen reduction, ammonia synthesis, among others. It has also been successful in providing insights into the design of novel surfaces for the catalysis of chemical and electrochemical reactions that are relevant in many industrial applications [1–6]. Consequently, computational modeling or simulation is now considered as the third pillar of scientific inquiry, after the theoretical and experimental methods as the first two.

Such ab initio (i.e., solely from fundamental physical principles, without empirical data) calculations entail solving the time-independent, non-relativistic Schrödinger equation:

$$\hat{H}\left|\Psi\left(\vec{r}_1,\vec{r}_2,\ldots\vec{r}_N,\vec{R}_1,\vec{R}_2,\ldots\vec{R}_M\right)\right\rangle = E\left|\Psi\left(\vec{r}_1,\vec{r}_2,\ldots\vec{r}_N,\vec{R}_1,\vec{R}_2,\ldots\vec{R}_M\right)\right\rangle. \tag{1.1}$$

© Springer Nature Singapore Pte Ltd. 2020

H. Kasai et al., *Hydrogen and Hydrogen-Containing Molecules on Metal Surfaces*, Springer Series in Surface Sciences 71,
https://doi.org/10.1007/978-981-15-6994-4_1

Here, the wavefunction Ψ is a function of real space coordinates $\vec{r}_i$ and $\vec{R}_I$, which denote respectively the positions of electron and nuclei in a system with 3(N + M) degrees of freedom. The Hamiltonian $\hat{H}$ for a system consisting of M nuclei and N electrons is:

$$\hat{H} = \hat{T}_{elec} + \hat{T}_{nuc} + \hat{V}_{elec-nuc} + \hat{V}_{elec-elec} + \hat{V}_{nuc-nuc}, \tag{1.2}$$

where the individual terms, expressed in SI units are:

$$\hat{T}_{elec} = -\frac{\hbar^2}{2m}\sum_{i=1}^{N}\nabla_i^2, \tag{1.3}$$

$$\hat{T}_{nuc} = -\frac{\hbar^2}{2}\sum_{I=1}^{M}\frac{1}{M_I}\nabla_I^2, \tag{1.4}$$

$$\hat{V}_{elec-nuc} = \frac{1}{4\pi\varepsilon_o}\sum_{i=1}^{N}\sum_{I=1}^{M}\frac{Z_I e}{r_{iI}}, \tag{1.5}$$

$$\hat{V}_{elec-elec} = \frac{1}{4\pi\varepsilon_o}\sum_{i<j}\frac{e^2}{r_{ij}} = \left(\frac{1}{2}\right)\frac{1}{4\pi\varepsilon_o}\sum_{i\neq j}\frac{e^2}{r_{ij}}, \tag{1.6}$$

$$\hat{V}_{nuc-nuc} = \frac{1}{4\pi\varepsilon_o}\sum_{I<J}\frac{Z_I Z_j}{R_{IJ}} = \left(\frac{1}{2}\right)\frac{1}{4\pi\varepsilon_o}\sum_{I\neq J}\frac{Z_I Z_j}{R_{IJ}}. \tag{1.7}$$

Here, I and J run over M nuclei while i and j denote the N electrons in the system. Also, ∇_I^2 represents the Laplacian with respect to nucleus I whose mass is M_I, while ∇_i^2 is the Laplacian with respect to the electron i, whose mass is m.

Obtaining an exact solution to the Schrödinger equation is extremely challenging (for almost all cases) except for the simplest ones. In this view, the Born-Oppenheimer approximation is implemented. Due to the large masses of the nuclei and their slower motion with respect to the electrons, the electrons can be considered as moving in field of fixed nuclei. Reactions in which this approximation is valid are commonly referred to as *electronically adiabatic*. Thus, the nuclear kinetic energy is approximately zero ($T_{nuc} \approx 0$) and their potential energy is merely a constant $\left(E_{nuc-nuc} = \frac{1}{4\pi\varepsilon_o}\sum_{I<J}\frac{Z_I Z_j}{R_{IJ}}\right)$. This reduces the electronic Hamiltonian to:

$$\hat{H}_{elec} = \hat{T}_{elec} + \hat{V}_{elec-nuc} + \hat{V}_{elec-elec}, \tag{1.8}$$

$$\hat{H}_{elec} = -\frac{\hbar^2}{2m}\sum_{i=1}^{N}\nabla_i^2 + \frac{1}{4\pi\varepsilon_o}\sum_{i=1}^{N}\sum_{I=1}^{M}\frac{Z_I e}{r_{iI}} + \frac{1}{4\pi\varepsilon_o}\sum_{i<j}\frac{e^2}{r_{ij}}. \tag{1.9}$$

The solution to the Schrödinger equation with $\hat{H}_{elec}$ is the electronic wavefunction Ψ_{elec} and electronic energy E_{elec}. The total energy E_{tot} is then the sum of E_{elec} and the constant nuclear repulsion term $E_{nuc-nuc} = \frac{1}{4\pi\varepsilon_o}\sum_{I<J}\frac{Z_I Z_J}{R_{IJ}}$. That is, $E_{tot} = E_{elec} + E_{nuc-nuc}$.

Total energy calculations can be performed within the framework of DFT, which states that if the ground state charge density is known for any given system of atoms, the external potential and the number of electrons can be known, and hence the full Hamiltonian $\hat{H}$. The density $\rho(\vec{r})$ thus implicitly determines all properties derivable from $\hat{H}$ through the solution of the Schrödinger equation. This conclusion arises from the *reductio ad absurdum* proof presented by Hohenberg and Kohn [7] in the 1960s, which is discussed as follows:

Consider two external potentials $V_{ext}(\vec{r})$ and $V'_{ext}(\vec{r})$ which give the same $\rho(\vec{r})$. There would be two Hamiltonians $\hat{H}$ and $\hat{H}'$ whose ground-state electron densities are the same although the normalized wave functions Ψ and Ψ' are different. Taking Ψ' as a trial wave function for the $\hat{H}$ problem,

$$E_o < \langle\Psi'|\hat{H}|\Psi'\rangle = \langle\Psi'|\hat{H}'|\Psi'\rangle + \langle\Psi'|\hat{H} - \hat{H}'|\Psi'\rangle, \tag{1.10}$$

$$E_o < E'_o + \int \rho(\vec{r})\left[V_{ext}(\vec{r}) - V'_{ext}(\vec{r})\right]d\vec{r}, \tag{1.11}$$

where E_o and E'_o are the ground-state energies for $\hat{H}$ and $\hat{H}'$ respectively. Similarly, taking Ψ as a trial function for the $\hat{H}'$ problem,

$$E'_o < \langle\Psi|\hat{H}'|\Psi\rangle = \langle\Psi|\hat{H}|\Psi\rangle + \langle\Psi|\hat{H}' - \hat{H}|\Psi\rangle, \tag{1.12}$$

$$E'_o < E_o + \int \rho(\vec{r})\left[V'_{ext}(\vec{r}) - V_{ext}(\vec{r})\right]d\vec{r}. \tag{1.13}$$

Adding 1.11 and 1.13 gives $E_o + E'_o < E'_o + E_o$, which is a contradiction. Therefore, there is a one-to-one mapping between the external potential $V_{ext}(\vec{r})$ and electron density $\rho(\vec{r})$. This is the first Hohenberg-Kohn theorem. The ground state energy can now be written as a unique functional of electron density:

$$E[\rho] = E_{elec-nuc}[\rho] + T[\rho] + E_{elec-elec}[\rho]. \tag{1.14}$$

The second Hohenberg-Kohn theorem establishes a variational principle in quantum mechanics which states that the energy functional assumes its minimum value if and only if the input density is the true ground state density. That is, for any trial density $\rho'(\vec{r})$ which satisfies the necessary boundary conditions such as $\rho'(\vec{r}) \geq 0$ and $\int \rho'(\vec{r})d\vec{r} = N$, the energy obtained $E[\rho']$ represents an upper bound to the true ground state energy E_o:

$$E_o \leq E[\rho'] = E_{elec-nuc}[\rho'] + T[\rho'] + E_{elec-elec}[\rho']. \tag{1.15}$$

So the ground-state energy of the system can be written as:

$$E = \min_{\rho(\vec{r})}(E_{elec-nuc}[\rho] + T[\rho] + E_{elec-elec}[\rho]), \tag{1.16}$$

$$E = \min_{\rho(\vec{r})}(\int \rho(\vec{r})V_{elec-nuc}(\vec{r})d\vec{r} + F[\rho]). \tag{1.17}$$

The functional $F[\rho]$ which contains the functional for kinetic energy and that for electron-electron interaction, $F[\rho] = T[\rho] + E_{elec-elec}[\rho]$, is the holy grail of DFT since its explicit form is unknown. The $E_{elec-elec}[\rho]$ can be decomposed into the classical and non-classical part:

$$E_{elec-elec}[\rho] = E_{classical}[\rho] + E_{non-classical}[\rho], \tag{1.18}$$

$$E_{elec-elec}[\rho] = \left(\frac{1}{2}\right)\frac{e^2}{4\pi\varepsilon_o}\iint \frac{\rho(\vec{r}_1)\rho(\vec{r}_2)}{r_{12}}d\vec{r}_1 d\vec{r}_2 + E_{non-classical}[\rho]. \tag{1.19}$$

The Thomas-Fermi model, which was introduced in 1927, provided the first expression that utilized electron density in solving the total energy of a system [8, 9]. However, its implementation is hindered by the inadequate representation of kinetic energy functional, the assumed non-interacting electrons in the system, and the neglected quantum effects. To solve this, Kohn and Sham suggested an approach which calculates the exact kinetic energy of a non-interacting single particle reference system $T_s[\rho]$ with the same electron density as the real interacting one.

$$T_s[\rho] = -\frac{\hbar^2}{2m}\sum_i^N \langle \Phi_i|\nabla^2|\Phi_i\rangle, \tag{1.20}$$

$$\rho_s = \sum_i^N |\Phi_i(\vec{r})|^2 = \rho(\vec{r}), \tag{1.21}$$

where Φ_i's are the orbitals of the non-interacting electrons. Since $T[\rho]$ and $T_s[\rho]$ are different, the functional $F[\rho]$ can be written as:

$$F[\rho] = T[\rho] + E_{elec-elec}[\rho] \tag{1.22}$$

$$F[\rho] = T[\rho] + (E_{classical}[\rho] + E_{non-classical}[\rho]) \tag{1.23}$$

$$F[\rho] = T[\rho] + (T_s[\rho] - T_s[\rho]) + (E_{classical}[\rho] + E_{non-classical}[\rho]) \tag{1.24}$$

$$F[\rho] = T_s[\rho] + E_{classical}[\rho] + \{(T[\rho] - T_s[\rho]) + E_{non-classical}[\rho]\}. \quad (1.25)$$

Enclosed in the brace is the so-called exchange-correlation energy, $E_{XC}[\rho]$ which contains everything that is unknown:

$$E_{XC}[\rho] = T[\rho] - T_s[\rho] + E_{non-classical}[\rho]. \quad (1.26)$$

So the energy functional can be written as:

$$\begin{aligned} E[\rho] = E_{elec-nuc}[\rho] + F[\rho] = E_{elec-nuc}[\rho] \\ + T_s[\rho] + E_{classical}[\rho] + E_{XC}[\rho] \end{aligned} \quad (1.27)$$

$$\begin{aligned} E[\rho] = & \int \rho(\vec{r}) V_{elec-nuc}(\vec{r}) d\vec{r} - \frac{\hbar^2}{2m} \sum_i^N \langle \Phi_i | \nabla^2 | \Phi_i \rangle \\ & + \left(\frac{1}{2}\right) \frac{e^2}{4\pi\varepsilon_o} \iint \frac{\rho(\vec{r}_1)\rho(\vec{r}_2)}{r_{12}} d\vec{r}_1 \vec{r}_2 \\ & + \int \rho(\vec{r}) V_{XC}(\vec{r}) d\vec{r} \end{aligned} \quad (1.28)$$

$$\begin{aligned} E[\rho] = & \frac{1}{4\pi\varepsilon_o} \sum_i^N \int \sum_I^M \frac{Z_I e}{r_{1I}} |\Phi_i(\vec{r}_1)|^2 d\vec{r}_1 - \frac{\hbar^2}{2m} \sum_i^N \langle \Phi_i | \nabla^2 | \Phi_i \rangle \\ & + \left(\frac{1}{2}\right) \frac{e^2}{4\pi\varepsilon_o} \sum_i^N \sum_j^M \iint |\Phi_i(\vec{r}_1)|^2 \frac{1}{r_{12}} |\Phi_j(\vec{r}_2)|^2 d\vec{r}_1 d\vec{r}_2 \\ & + \sum_i^N \int |\Phi_i(\vec{r}_1)|^2 V_{XC}(\vec{r}) d\vec{r}_1. \end{aligned} \quad (1.29)$$

Changing the variables: $\vec{r}_1 \rightarrow \vec{r}$ and $\vec{r}_2 \rightarrow \vec{r}'$ and applying the variational principle under the usual constraint $\langle \Phi_i | \Phi_i \rangle = \delta_{ij}$, the resulting equation is:

$$\begin{aligned} & \left(-\frac{\hbar^2}{2m} \nabla^2 + \left[\frac{e^2}{4\pi\varepsilon_o} \int \frac{\rho(\vec{r}')}{|\vec{r} - \vec{r}'|} d\vec{r}' + V_{XC}(\vec{r}) \right. \right. \\ & \left. \left. + \frac{e^2}{4\pi\varepsilon_o} \sum_I^M \frac{Z_I}{|\vec{r} - \vec{r}_I|} \right] \right) \Phi_i(\vec{r}) = \varepsilon_i \Phi_i(\vec{r}), \end{aligned} \quad (1.30)$$

where the terms enclosed in the bracket comprise the effective potential:

$$V_{eff}(\vec{r}) = \frac{e^2}{4\pi\varepsilon_o} \int \frac{\rho(\vec{r}')}{|\vec{r} - \vec{r}'|} d\vec{r}' + V_{XC}(\vec{r})$$

$$+\frac{e^2}{4\pi\varepsilon_o}\sum_I^M \frac{Z_I}{|\vec{r}-\vec{r}_I|}. \quad (1.31)$$

The first and last terms in 1.31 are the external potential V_{ext} and Hartree potential V_H, respectively. Thus, 1.30 can be written as:

$$\left(-\frac{\hbar^2}{2m}\nabla^2 + V_{eff}(\vec{r})\right)\Phi_i = \varepsilon_i\Phi_i(\vec{r}). \quad (1.32)$$

Equations 1.21, 1.31 and 1.32 are the celebrated Kohn-Sham (KS) equations. For clarity, these equations are summarized below:

$$\rho(\vec{r}) = \sum_i^N |\Phi_i(\vec{r})|^2, \quad (1.33)$$

$$\left(-\frac{\hbar^2}{2m}\nabla^2 + V_{eff}(\vec{r})\right)\Phi_i = \varepsilon_i\Phi_i(\vec{r}), \quad (1.34)$$

$$V_{eff}(\vec{r}) = V_{ext}(\vec{r}) + V_H(\vec{r}) + V_{XC}(\vec{r}). \quad (1.35)$$

Since $V_H(\vec{r})$ and $V_{XC}(\vec{r})$ depend on $\rho(\vec{r})$, which depends on $\Phi_i(\vec{r})$, which in turn depends $V_{eff}(\vec{r})$, the problem of solving the KS equations in a nonlinear one. As shown in Fig. 1.1, the computational approach usually starts with an initial guess for $\rho(\vec{r})$, calculate the corresponding $V_{eff}(\vec{r})$, solve the differential equation for $\Phi_i(\vec{r})$, and then calculate the new $\rho(\vec{r})$. This process is repeated until it converges—a procedure called "self-consistent field cycle".

For an arbitrary $\rho(\vec{r})$, the exact expression for $E_{XC}[\rho]$ is not known. Hence, the reliability of any practical implementation of density functional theory crucially depends on the accuracy of the expression for the exchange-correlation energy functional. The simplest and historically the most important approximation is the *Local Density Approximation* (LDA), wherein at any point $\vec{r}$, the local exchange-correlation energy of the homogeneous electron gas is used for the corresponding electron density, ignoring the general nonlocality of the true exchange-correlation energy:

$$E_{XC}^{LDA}[\rho] = \int \rho(\vec{r})\varepsilon_{XC}(\rho(\vec{r}))d\vec{r}. \quad (1.36)$$

Here, $\varepsilon_{XC}(\rho(\vec{r}))$ is the exchange-correlation energy per particle of a uniform electron gas of density $\rho(\vec{r})$ and can be further decomposed into exchange and correlation contributions:

$$\varepsilon_{XC}(\rho(\vec{r})) = \varepsilon_X(\rho(\vec{r})) + \varepsilon_C(\rho(\vec{r})). \quad (1.37)$$

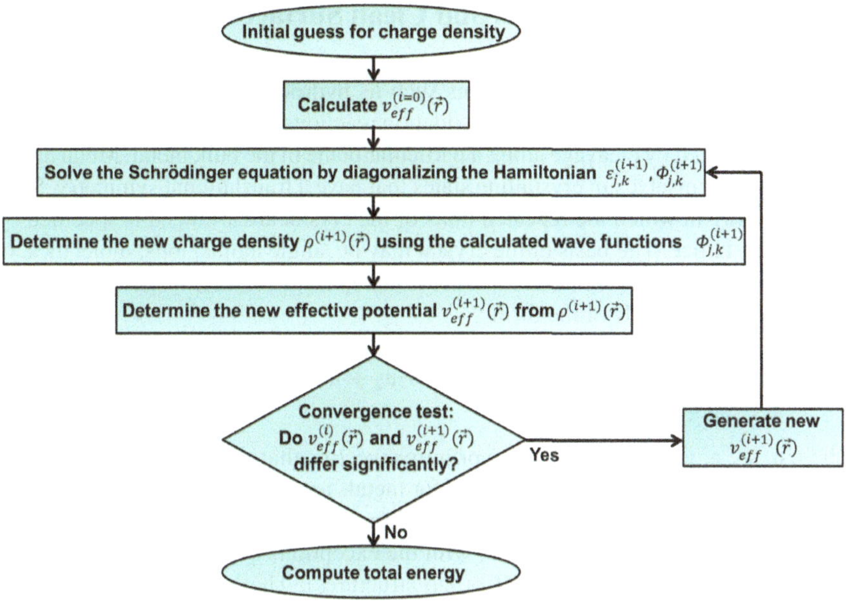

Fig. 1.1 Self-consistent calculation of the ground state electron density using Kohn-Sham equations

Since any real system is spatially inhomogeneous, i.e., it has spatially varying density $\rho(\vec{r})$, it would clearly be important to include information on the rate of this variation in the functional. Hence, it was a major breakthrough in the early eighties when it was realized that one could include general functions of $\rho(\vec{r})$ and $\nabla\rho(\vec{r})$ in the functional. Such functional of the general form

$$E_{XC}^{GGA}[\rho] = \int \rho(\vec{r}) f(\rho(\vec{r}), \nabla\rho(\vec{r})) d\vec{r}. \tag{1.38}$$

has become known as *Generalized-Gradient Approximation* (GGA) [10–12]. The most popular and reliable GGAs are PBE (denoting the functional proposed in 1996 by Perdew et al. [10]) in physics and BLYP (combination of Becke's 1988 exchange functional [13] with the 1988 correlation functional of Lee et al. [14]) in chemistry.

1.2 Crystalline Structures and Clean Surfaces

Many industrially significant reactions such as hydrocarbon reforming, hydrogen and oxygen evolution reactions, and oxidation reactions occur on the metal surfaces, which are created by cleavage along a particular plane in the bulk metal. Metal atoms can arrange themselves in crystalline states that have a translational symmetry. Such periodic array in which the repeated units of the crystal are arranged is specified by the Bravais lattice. The basis of the crystal may be single atoms, ions, molecules, or polymer strings of a solid. A three-dimensional Bravais lattice consists of all points with position vectors R of the form

$$\mathbf{R} = n_1\mathbf{a}_1 + n_2\mathbf{a}_2 + n_3\mathbf{a}_3 \tag{1.39}$$

where n_i are any integers and $\mathbf{a}_i$ are primitive vectors that span the lattice.

The most common crystal structures for metals are the face-centered cubic (fcc), body-centered cubic (bcc), and hexagonal close-packed (hcp) structures. Most transition metals take the fcc structures, with the exceptions of Fe, Mo, and W (taking the bcc structures), and Co and Ru (hcp structures). The lattice vectors for these structures are

fcc structure:

$$\mathbf{a}_1 = \frac{a}{2}(\hat{\mathbf{y}} + \hat{\mathbf{z}}),\quad \mathbf{a}_2 = \frac{a}{2}(\hat{\mathbf{z}} + \hat{\mathbf{x}}),\quad \mathbf{a}_3 = \frac{a}{2}(\hat{\mathbf{x}} + \hat{\mathbf{y}}) \tag{1.40}$$

bcc structure:

$$\mathbf{a}_1 = a\hat{\mathbf{x}},\quad \mathbf{a}_2 = a\hat{\mathbf{y}},\quad \mathbf{a}_3 = \frac{a}{2}(\hat{\mathbf{x}} + \hat{\mathbf{y}} + \hat{\mathbf{z}}), \tag{1.41}$$

where a is the lattice constant and each unit has a single atom.

For hcp structure, the lattice vectors are:

$$\mathbf{a}_1 = a\hat{\mathbf{x}},\quad \mathbf{a}_2 = \frac{a}{2}\hat{\mathbf{x}} + \frac{\sqrt{3}}{2}\hat{\mathbf{y}},\quad \mathbf{a}_3 = c\hat{\mathbf{z}}. \tag{1.42}$$

In this structure, the first two lattice vectors generate a triangle in the xy plane, while the third vector places the nets of triangles a distance *c* above the first one. The hcp structure is generated by interpenetrating simple hexagonal Bravais lattices, displaced from one another by $\mathbf{a}_1/3 + \mathbf{a}_2/3 + \mathbf{a}_3/2$. For an ideal hcp structure, the c/a ratio is $\sqrt{(8/3)}$ or 1.633. These structures are shown in Fig. 1.2.

Another important structure is the diamond structure, which is assumed by some group IV elements such as C, Si, and Ge. It consists of two interpenetrating fcc Bravais lattices, displaced along the body diagonal of the cubic cell by one quarter of the length of the diagonal. It can be regarded as fcc lattice with two-point basis at

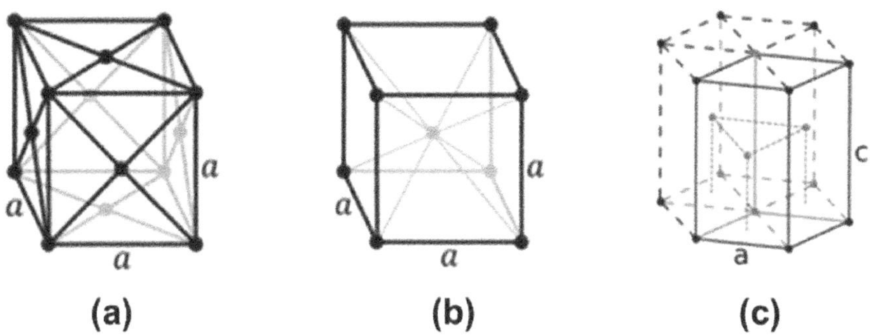

Fig. 1.2 The most common crystal structures for metals: **a** face centered cubic (fcc), **b** body centered cubic (bcc), and **c** hexagonal close-packed (hcp) structures. Reprinted with permission from [15]. Copyright 2008 by Wiley

$$0, \frac{a}{4}(\hat{\mathbf{x}} + \hat{\mathbf{y}} + \hat{\mathbf{z}}) \tag{1.43}$$

Surfaces can be constructed by cleavage along any arbitrary plane in a perfect crystal. The directions in the lattice that define a particular plane are indicated by *Miller indices*. A plane of atoms is uniquely defined by the direction that is normal to the plane. In real space, Miller indices can be defined as inverse of intercepts expressed as integers (by multiplying the inverse of intercepts by their common denominator) along the lattice vectors. The (hkl) notation denotes a plane that intercepts the three points $\mathbf{a}_1/h$, $\mathbf{a}_2/k$, and $\mathbf{a}_3/l$. A set of all planes equivalent to (hkl) by the symmetry of the lattice is denoted by {hkl}. In the context of crystal directions (not planes), the notation [hkl] is used to denote a direction in the basis of the direct lattice vectors. Some common crystal planes with the corresponding Miller indices are shown in Fig. 1.3.

The simplest and flattest of the fundamental planes are the *low index planes*. The term "low" is used because the Miller indices for such planes are small numbers, either 0 or 1. For example, the low index planes in the fcc structure are (100), (110), and (111) surfaces. On the other hand, *high index planes* (h, k, or l > 1) have open structures that can expose the sublayer atoms. Cutting the crystal at high index planes can lead to *vicinal surfaces*, where step-like discontinuities are introduced into the surface structure. A vicinal surface is composed of a series of terraces and steps (and might include kinks), hence, they are also known as *stepped surfaces*. One example is the fcc (211) surface, which is composed of three-atom-wide terraces with (111) facet structure and one-atom step with (100) character.

Because stepped surfaces have both terraces and steps, they provide an interesting surface heterogeneity that directly affects the properties of the materials. The electrons of the atoms at the step edge site tend to smooth out the discontinuity at the step in a process known as *Smoluchowski smoothing*. This has implications, for instance, in the diffusion of adsorbates over steps. Moreover, since the electronic properties of steps differs from that of terraces, the chemical reactivity is different as well. For

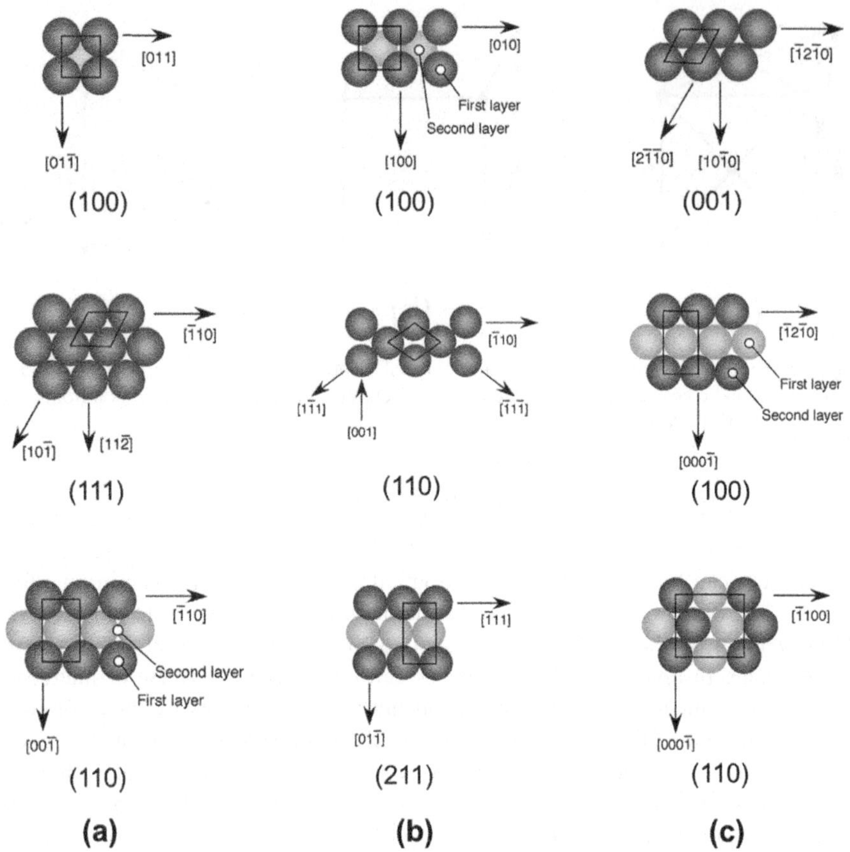

Fig. 1.3 Representations of some common crystal planes for **a** fcc, **b** bcc, and **c** hcp structures, with the corresponding Miller indices. Reprinted with permission from [15]. Copyright 2008 by Wiley

instance, a study on the electronic states of Ru by Arevalo et al. showed that with respect to the bulk case, the d band of Ru is more forward-shifted (skewed) to the Fermi level for the case of the step-edge atom of the stepped surface compared to the terrace atom of the flat surface. This indicates an enhanced reactivity of the stepped surface compared to the low-index flat surface [16].

When a particle adsorbs on surfaces, the number of surface atoms that binds directly to the adsorbate could be used to describe the different adsorption sites on the surface. Focusing on the low index planes of the fcc crystals, different high symmetry sites can be identified: The (111) surface has one-fold (top site), two-fold (bridge site), and three-fold (fcc and hcp hollow sites) coordinated sites. The (100) surface has one-fold (top site), two-fold (bridge site), and four-fold (hollow site between four atoms) coordinated sites. Also, the (110) surface displays two different two-fold sites: a long bridge site between two atoms on adjacent rows and a short

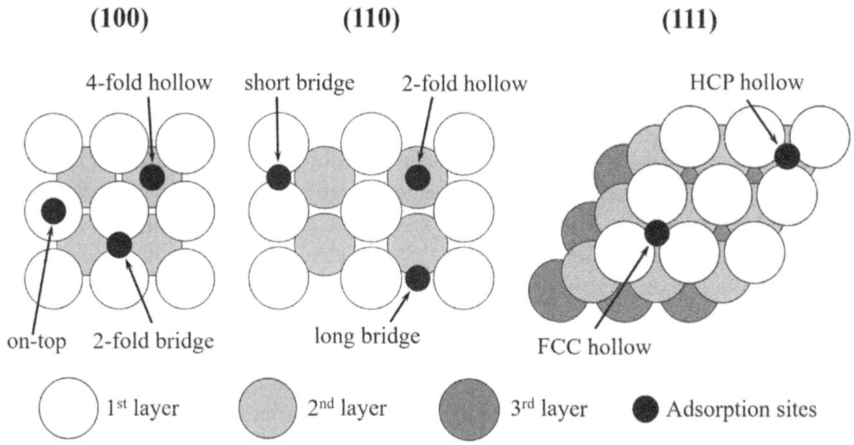

Fig. 1.4 Different possible adsorption sites on the (100), (111), and (110) surfaces of an fcc crystal. Reprinted with permission from [15]. Copyright 2008 by Wiley

bridge site between two atoms in the same row. These adsorption sites are shown in Fig. 1.4.

While many pure metals are active in catalyzing a variety of surface reactions, the issues of catalyst activity, selectivity, poisoning and other desired properties necessitate the design of new materials for the desired surface reactions. One approach is to mix two or more metals to achieve unique material properties that fit the desired requirements of particular reactions. For example, pure Au is commonly used to catalyze the oxidation of borohydride for fuel cell application. However, high overpotential (thus, large energy loss) is necessary to achieve an appreciable rate of borohydride on Au. Interestingly, researchers have found theoretically and experimentally [17, 18] that alloying Au with the third row transition metals can promote the oxidation of borohydride at lower overpotential than pure Au. Similarly, other researchers have found that alloying Pt with Ni can effectively enhance the activity of Pt towards oxygen reduction reaction, in addition to reducing the cost of the catalyst [19].

The class of such material that is composed of a mixture of metals is called an *alloy*. It is a solid solution where metals are randomly dissolved in the solution; that is, the solute will tend to be randomly distributed in the solvent. If the surface energy of one component of the alloy is significantly lower than that of the other components, the component with low surface energy will preferentially segregate on the surface. This results in the enrichment of the surface with one of the elements of the alloys. An example is the segregation of Pt on third row transition metals which find applications for the dissociation of oxygen gas [20]. On the other hand, some mixtures of metals form an *intermetallic compound*, which has a definite and uniform stoichiometry and unit cell. Some examples of intermetallic compounds are Cu_3Au, Ni_3Al, Pt_3Ni, and NiAl.

1.3 Chemisorption

The adsorption of molecules on surfaces is a fundamental concept in many phenomena such as catalysis, corrosion, film growth, among others. In general, chemical reactions on surfaces can be understood as a combination of elementary processes such as adsorption, dissociation, diffusion, association, desorption, and scattering. Each elementary process constitutes a change in the electronic and nuclear states of the system. In the adsorption process, the adsorbates form bonds with atoms of the surface, and are stabilized with the release of energy. The released energy is transferred to the surface or another adsorbate and causes the excitation of other electrons or might result to nuclear motion. When the excited adsorbate has an energy that exceeds the activation energy for breaking its bond with the surface, the adsorbate can desorb from the surface. If the energy of the adsorbate is less than its adsorption energy but exceeds the activation barrier to form bonds with other atoms on the surface, the adsorbate can diffuse over the surface. If the adsorption energy is larger than thermal fluctuation, a stable adsorption state is obtained in the thermal equilibrium state; and if the opposite is true, the adsorption and desorption processes are repeated. This illustrates that the adsorption process is the basic elementary step in the reaction of atoms and molecules approaching the surface and is an important process in the characterization of the overall reaction mechanism.

There are two types of adsorption which can higly depend on the distance of the adsorbate from the surface. In the *chemisorption* type of adsorption, the energy gain from orbital hybridization between the adsorption and surface electronic states provides stability. Because of the electron transfer between the adsorbate and the surface, polarization may occur due to the difference in electronegativities of atoms of the adsorbate and of the surface. When the electron transfer is large, the adsorbate is ionized, and the characteristic of coupling by electrostatic attraction also appears. In the chemisorption process, bonds in the adsorbate can break as it forms new bonds with the surface, resulting in its dissociation (dissociative adsorption). The reverse process is the associative desorption. On the other hand, the adsorption is of *physisorption* type if the adsorbate is relatively far from the surface and the adsorption process does not involve orbital hybridization or charge transfer; that is, the energy gain is by virture of the long-range intermolecular forces. This type of adsorption will be discussed in detail in the next section.

The interaction between the adsorbate and the surface at the vicinity of the bonding distance in chemisorption can be roughly described by many-body and local Coulomb interactions acting between electrons and nuclei in the vicinity of the bond. As the adsorbate approaches the surface from infinity, it can be electrostatically attracted to the surface (detailed mechanism will be explained in the next section). When the distance of separation between the surface atoms and adsorbates is within their atomic radii, the Pauli exclusion principle dictates a repulsive Coulomb interaction between closed electron shells. As the adsorbate moves even closer to the surface, bonds are formed by the hybridization of adsorbate and surface states, with a corresponding energy gain. This process may be accompanied by dissociation or a change in the

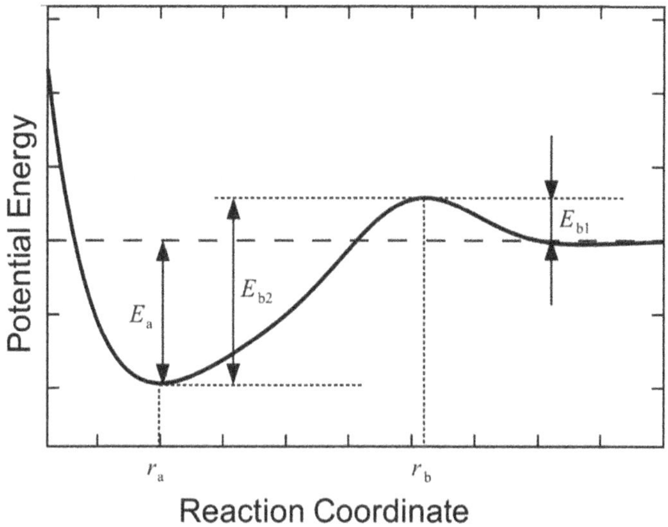

Fig. 1.5 Conceptual diagram of potential energy curve for the chemisorption process. The dashed horizontal lines indicate the energy at the limit of an infinite distance of separation between the adsorbate and the surface. r_a and r_b are the distance coordinates of the adsorbate at the stable adsorption site and at the transition state, respectively. E_a denotes the adsorption energy. E_{b1} and E_{b2} represent activation barriers in the adsorption and desorption processes, respectively. Adapted with permission from [21]. Copyright 2013 by Asakura Publishing

structure of the adsorbate. As the adsorbate moves closer to the surface, a repulsion between inner-shell electrons and nuclei occurs. As a result, the potential energy $U(r)$ exhibits an energy barrier at a distance r_b slightly away from the adsorption distance r_a (see Fig. 1.5).

For molecules to be adsorbed on the surface, the molecules must have kinetic energies that exceed the activation barrier of $E_{\mathrm{b}1} = U(r_\mathrm{b}) - U(\infty)$. Conversely, for an adsorbed molecule to desorb from the surface, the adsorbate must have an energy greater than the activation barrier for desorption given by $E_{\mathrm{b}2} = U(r_\mathrm{b}) - U(r_\mathrm{a})$. Some systems can overcome the adsorption barrier by considering the contribution from other degrees of freedom such as molecular rotation. Assuming reversibility, the desorption activation barrier for this system is simply the adsorption energy $E_\mathrm{a} = U(\infty) - U(r_\mathrm{a})$. Basically, the chemisorption of molecules ceases as they form an adlayer, thereby blocking the bonding sites on the surface (the surface energy is offset by the adsorption energy). The surface to molecule adsorption distance is close to the sum of the covalent bond radii of the surface and adsorbate atoms, and is about a few angstroms. The adsorption energy of the molecule in such case is about 0.1–10.0 eV. The distance dependence of the potential energy near the adsorption distance is scaled by the overlap of electron wave functions between atoms. Since the tail of a general wave function decays exponentially, the potential energy of the attractive part also decays exponentially with distance. *Morse potential* is a simple expression of potential energy reflecting this feature:

$$U(r) = E\left[e^{-2a(r-r_a)} - 2e^{-a(r-r_a)}\right]. \quad (1.44)$$

Here, the second term in parentheses represents the attractive part. The first term is a repulsive term introduced for computational simplicity. The parameter a gives the spread of the potential well. Morse potential can be described as harmonic vibration with well-defined anharmonic terms; so it is useful to analyze the vibrational state near the adsorption distance. It should be noted that the asymptotic behavior at a finite potential energy value for $r \to +0$ is inappropriate (i.e., the potential energy should be $U(0) \to \infty$). This is especially true for the case of on-top adsorption (adsorption onto a surface atom). While phenomenological potential-based analysis is convenient in the study of dynamic processes, first-principles calculation based on DFT is mainly used for the analysis of stable adsorption states. Though DFT-based calculations remain to pose unsolved challenges such as the reliability of the exchange-correlation functional, accuracy of the unoccupied states above the Fermi energy, and estimation of band gaps, it has still been proven to provide very high reliability for the analysis of the electronic ground state of chemical substances. This makes DFT-based first principles calculations a robust method in the analysis of chemisorption in the electronic ground state.

1.4 Physisorption

Intermolecular forces are almost present in between all substances consisting of nuclei and electrons, such as atoms, molecules, crystals, and others. It is a major attractive interaction between substances that are far away from each other. Surfaces with high degree of condensation (such as solid surfaces and molecules) and fluid particles with low degree of condensation (atoms and molecules that are not chemically bound to each other) are combined at a distance where the influence between attractive intermolecular force and repulsive force between electron clouds are balanced. This state is a *physisorption* state. In a broader sense, intermolecular forces include the **van der Waals force**, as well as the interionic Coulomb interaction and hydrogen bond. However, in a physisorption state, the interaction is usually by van der Waals force. In principle, there is no activation barrier for the process of physisorption because intermolecular forces act exclusively as attractive forces (unless they are affected by homopolar repulsion and do not lead to adsorption).

The van der Waals force is a common term to describe the attractive interaction between molecules and in noble gases. In molecules composed of heteroatoms, electrons gather at highly electronegative atoms and behave as polar molecules; that is, permanent multipoles. Considering the case of a permanent dipole for simplicity, the potential energy between two stationary dipoles is inversely proportional to the cube of the distance between them. Since the binding energy due to this interaction is very small (e.g., as in a general liquid), the heat fluctuation keeps the stationary state and the molecule rotates almost freely. When taking time averages and rotational degrees of freedom into account, the effective potential energy is inversely proportional to the sixth power of the distance. This interaction is called dipolar-dipolar interaction or *Keesom interaction*.

Nonpolar neutral particles (molecules or noble gas atoms) located in the vicinity of polar molecules can be polarized by the electric field generated by the polar molecules. An attractive force also acts between these polarized neutral particles and the polar molecule. Because the strength of the induced electric field is inversely proportional to the cube of the distance, and the induced electric field interacts with the original polar molecule, the potential energy between the polar molecule and the nonpolar neutral particle is inversely proportional to the sixth power of the distance. This interaction is called the induced dipole interaction or *Debye interaction*. Between nonpolar neutral particles, a temporary electric multipole due to the electronic motion around the nucleus also creates an attractive force called dispersive force. Considering the classical electron motion for simplicity, a single nonpolar neutral particle (molecule or noble gas atom) seen from a distance away from its atomic radius is approximately neutral, but is polarized at a given instantaneous time. Therefore, assuming that electrons move periodically so that polarization is oriented between approaching nonpolar neutral particles, interactions between the neutral particles will occur. The potential energy between dipoles at a particular time is inversely proportional to the cube of the distance, similar to the case of permanent dipoles. Because electrons move around the nucleus in three dimensions, the time-average potential energy that takes into account their degrees of freedom is inversely proportional to the sixth power of the distance. The spontaneous orientation of the polarization assumed above can not be explained within the framework of classical mechanics. However, considering the quantum mechanical fluctuations of electronic coordinates, the same phenomenon can be considered as spontaneous polarization. The force acting between spontaneous polarizations caused by quantum mechanical fluctuation is called *London dispersion force*. This type of interaction, in general, is present in all kinds of atoms and molecules, and in many cases has a larger contribution compared to the Keesom and Debye interactions. Thus, it can be assumed that the van der Waals force is dominated by the London dispersion force.

As a simple expression of potential energy by van der Waals force, an approximation called *Lennard-Jones potential* is often used:

$$U(r) = 4 \in \left[\left(\frac{\sigma}{r} \right)^{12} - \left(\frac{\sigma}{r} \right)^{6} \right]. \tag{1.45}$$

Here, the attractive interaction term, that is inversely proportional to the sixth power of the distance r, represents the van der Waals potential due to the dipolar interaction. The repulsive interaction term, which is inversely proportional to the twelfth power of the distance, represents the repulsive force between the electron clouds. The Lennard-Jones potential is based on isotropic two-body interaction and can not be applied to general substances in which anisotropy exists. The coefficients $\in$ and σ are unqiue for each element (assumed to be isolated), and methods that treat them as effective interatomic interactions are generally successful for many compounds. In this case, $\in$ and σ are fitted to reproduce the experimental interatomic distances and bond energies. Because the only interaction between noble gases is the van der Waals force, the correspondence with the experiment is clearly established. For elements that form a simple diatomic molecule, reference is made based on experiments in molecular crystals (in such molecules, the contribution of the quadrupole term is large, but correction can be made by dipolar interaction by correcting the coefficient). In the case of carbon, since the interlayer coupling of graphite is due to van der Waals force, the values of the constants could be obtained from experimental results. If other elements, i.e. single crystals, have covalent or metallic bonds, it is difficult to associate with experiments, and there is no widely accepted method yet.

In first-principles calculations based on DFT, the standard approximation methods (local density approximation and generalized density gradient approximation) can only describe local interactions. It is therefore impossible to derive van der Waals forces from these methods. One solution is to add van der Waals potential between atoms i and j as a semi-empirical correction term, such as in *Grimme's DFT-D2 method*:

$$\mathrm{U_d(r)} = -\mathrm{s}\frac{\sqrt{\mathrm{C_iC_j}}}{\mathrm{r}^6}\frac{1}{1 - e^{-\mathrm{d}[\mathrm{r}/(\mathrm{R_i}+\mathrm{R_j})-1]}}, \tag{1.46}$$

where s is the overall scaling factor, d is the damping factor, R_i is the van der Waals radius of the atom, and C_i is the dispersion factor of the atom i. The dispersion coefficients (J-nm^6-mol^{-1}) are calculated using the ionization potential I_i (atomic unit), dipolar polarizability α_i (atomic unit), and rare gas atomic number N:

$$\mathrm{C_i} = 0.05\,\mathrm{NI_i}\alpha_\mathrm{i}. \tag{1.47}$$

When a rare gas atom or a small molecule is physisorbed on the surface, the van der Waals force between the atom or molecule and the surface atoms in its vicinity results to an attractive force. Similar to the bonds between the molecules, it only has a bond energy with a thermal fluctuation of ~ 0.01 eV. However, when carbon-based macromolecules such as carbon nanotubes are physisorbed on the surface, the energy required to separate them increases in proportion to the number of atoms in the contact interface, making the adsorption very strong. In principle, adsorption and desorption can occur reversibly, and the amount of adsorption is generally determined by macroscopic statistical mechanical quantities such as pressure and temperature. However, in some cases, reactions specific to physisorption such as ortho-para conversion occur in the state where hydrogen molecules are physisorbed on the solid surface. An adsorbate could be physisorbed before it overcomes the activation barrier. Then eventually, it could be chemisorbed on the surface. The physisorbed adsorbate exerts an intermolecular force on another particle as it approaches, so a multi-layer adsorption is possible.

When diatomic molecules such as H_2, NO, CO, etc. are physisorbed, orienting the molecular axis perpendicular to the surface tends to be slightly more stable than the surface parallel orientation. For the case of surface normal orientation, weak hybridization between the antibonding orbital of the molecule and the surface electron orbital occurs, making the Pauli repulsion between the closed shell electron orbital of the molecule and the surface reduce by half compared to the surface parallel orientation.

1.5 Dissociative Adsorption and Associative Desorption

In this section, the interaction between diatomic molecules such as hydrogen and oxygen and solid surfaces such as copper will be used as an example to explain the mechanism of dissociative adsorption and associative desorption. Elementary processes such as adsorption, dissociation, association, and desorption can be microscopically described as the change in the motion state of electrons and nuclei. Because the mass ratio of electrons to protons (or neutrons) is 1:1836, the time scales of their motions are also different (although there are some variations depending on the process). For this reason, the adiabatic approximation (Born-Oppenheimer Approximation), which assumes that the nucleus is almost stationary with respect to the electrons, could serve as an initial approximation to treat the interacting electrons and nuclei. As mentioned in the first section of this Chapter, when this approximation is used, the Hamiltonian is:

$$H_e\left(\vec{R}_1,\ldots,\vec{R}_{N_a};\vec{r}_1,\ldots,\vec{r}_{N_e}\right)=\sum_{i=1}^{N_e}\left[-\frac{\vec{p}_i^{\,2}}{2m_e}+\frac{e^2}{4\pi\epsilon_0}\sum_{m=1}^{N_a}\frac{Z_m}{\left|\vec{R}_m-\vec{r}_i\right|}+\frac{e^2}{4\pi\epsilon_0}\frac{1}{\left|\vec{r}_i-\vec{r}_j\right|}\right]+\frac{e^2}{4\pi\epsilon_0}\frac{Z_mZ_n}{\left|\vec{R}_m-\vec{R}_n\right|}. \quad (1.48)$$

For the nuclear systems:

$$H_a\left(\vec{R}_1,\ldots,\vec{R}_{N_a}\right)=\sum_{m=1}^{N_a}-\frac{\vec{P}_m^{\,2}}{2M_m}+V_k\left(\vec{R}_1,\ldots,\vec{R}_{N_a}\right) \quad (1.49)$$

$$V_k\left(\vec{R}_1,\ldots,\vec{R}_{N_a}\right)=\frac{\psi_k\left|H_a\left(\vec{R}_1,\ldots,\vec{R}_{N_a}\right)\right|\psi_k}{\psi_k|\psi_k} \quad (1.50)$$

in which, the matrix elements of the operator A is defined by:

$$\left\langle \psi_{k'} | A\left(\vec{R}_1, \ldots, \vec{R}_{N_a}\right) | \psi_k \right\rangle = \int d\vec{r}_1 \ldots \int d\vec{r}_{N_e} \psi^*_{k'}\left(\vec{R}_1, \ldots, \vec{R}_{N_a}; \vec{r}_1, \ldots, \vec{r}_{N_e}\right)$$
$$\times A\left(\vec{R}_1, \ldots, \vec{R}_{N_a}; \vec{r}_1, \ldots, \vec{r}_{N_e}\right)$$
$$\times \psi_k\left(\vec{R}_1, \ldots, \vec{R}_{N_a}; \vec{r}_1, \ldots, \vec{r}_{N_e}\right). \quad (1.51)$$

Here, the N_e is the total number of the electrons, N_a is the total number of nuclei, $\vec{r}_i$ and $\vec{R}_m$ are the positions of the ith electron and mth nuclei, $\vec{p}_i$. and $\vec{P}_m$ are the momentum of the ith electron and mth nuclei, respectively. Z_m is the atomic number of the mth nucleus. M_m and m_e are the masses of the mth nucleus and the electron, respectively. ψ_k is the eigenfunction for the eigenstate k of $H_e\left(\vec{R}_1, \ldots, \vec{R}_{N_a}; \vec{r}_1, \ldots, \vec{r}_{N_e}\right)$ and is a multielectron wave function represented by the Slater determinant. The intrinsic energy $V_k\left(\vec{R}_1, \ldots, \vec{R}_{N_a}\right)$ for the state k gives the adiabatic potential energy for nuclear motion.

Dissociative adsorption and associative desorption are opposite processes. If they proceed in the low energy region where electronic excitation does not occur, they can be treated using the same adiabatic potential energy surface. The difference between the two is the starting condition. In dissociative adsorption, the main starting condition is given by the state in which the molecule has momentum at a certain distance from the surface and the corresponding kinetic energy. In associative desorption, the process starts in a configuration where the atoms are present or adsorbed on the surface. External factors may provide additional kinetic energy to the atoms for them to cross the activation barrier and form molecule. These external factors may be due to increase in temperature, electronic stimulation, light stimulation, etc.

Consider the problem of a hydrogen gas molecule moving at low speed above a metal surface kept at absolute zero. For simplicity, assume that the metal atoms are sufficiently heavier than hydrogen atoms, and that the coordinates are fixed before, during, and after the reaction. In other words, the degrees of freedom of nuclear coordinates system are only for the hydrogen coordinates $\left(\vec{R}_{H1}, \vec{R}_{H2}\right)$. The electron system is in the ground state $k = 0$ with energy $V_0\left(\vec{R}_{H1}, \vec{R}_{H2}\right)$ in the atomic configuration given by each $\left(\vec{R}_{H1}, \vec{R}_{H2}\right)$ pair. The Hamiltonian for nuclear systems is simplified as follows:

$$H_a\left(\vec{R}_{H1}, \vec{R}_{H2}\right) = -\frac{\vec{P}^2_{H1}}{2M_{H1}} - \frac{\vec{P}^2_{H2}}{2M_{H2}} + V_0\left(\vec{R}_{H1}, \vec{R}_{H2}\right). \quad (1.52)$$

Here, M_{H1} and M_{H2} are the masses of hydrogen nuclei 1 and 2, which are separated to distinguish the different isotopes of hydrogen (light hydrogen, deuterium, and tritium). The adiabatic potential energy $V_0\left(\vec{R}_{H1}, \vec{R}_{H2}\right)$ does not depend on the nuclear mass, is symmetrical with respect to the exchange of coordinates, and has six degrees of freedom.

When the incident energy is low, the hydrogen nuclei move or relax to a low energy configuration. As a result, hydrogen nuclear motion proceeds along the reaction path that follows the valley of the adiabatic potential energy. For the nuclei motion, the center-of-mass coordinate $(X, Y, Z; r, \theta, \phi)$ could be used instead of the coordinates $\vec{R}_{\mathrm{H1}} = (x_1, y_1, z_1)$ and $\vec{R}_{\mathrm{H2}} = (x_2, y_2, z_2)$. Here, $(X, Y, Z) = \left(\vec{R}_{\mathrm{H1}} + \vec{R}_{\mathrm{H2}}\right)/2$, (r, θ, ϕ) is the spherical coordinate which represents the internal degree of freedom $\vec{R}_{\mathrm{H1}} - \vec{R}_{\mathrm{H2}}$ with respect to the cartesian coordinate in which the x and y axes are parallel to the surface while the z axis is normal to the surface. The form of $\vec{R}_{\mathrm{H1}} - \vec{R}_{\mathrm{H2}}$ can be written as follows:

$$\vec{\mathrm{R}}_{H1} - \vec{\mathrm{R}}_{H2} = (\mathrm{r} \sin\theta \cos\phi, \mathrm{r} \sin\theta \sin\phi, \mathrm{r} \cos\theta). \tag{1.53}$$

Here, $\left|\vec{R}_{\mathrm{H1}} - \vec{R}_{\mathrm{H2}}\right|$ is the internuclear distance, θ is the angle between the molecular axis and the z axis, and ϕ is the angle between the molecular axis projected on the surface and the x axis (see Fig. 1.6).

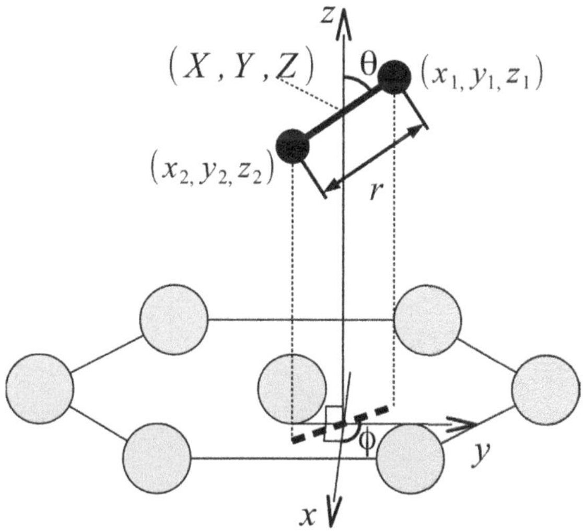

Fig. 1.6 Schematic diagram of space coordinates of a diatomic molecule placed on a solid surface. Gray and black circles represent the surface and atoms of the adsorbed molecule, respectively. The straight line connecting the black circles indicates the molecular axis. The straight line connecting gray circles indicates the topmost layer of the surface. The thick dashed line indicates the molecular axis projected to the topmost layer. The solid line with an arrow in a single direction indicates the direction of the Cartesian coordinate axis (the intersection point is projected into the midpoint of the molecular axis). (x_1, y_1, z_1) and (x_2, y_2, z_2) represent the coordinates of atoms 1 and 2 of the molecule, respectively. (X, Y, Z) represents the coordinates of the midpoint of the molecular axis. The variable r is the interatomic distance of the adsorbed molecules, θ is the angle between the molecular axis and the z-axis, and ϕ is the angle between the molecular axis projected onto the surface and the x-axis. Reprinted with permission from [21]. Copyright 2013 by Asakura Publishing

The relevant variables in the dissociative adsorption of diatomic molecules are r and Z. When the molecule is far away from the surface, the adiabatic potential energy depends mainly on r and Z because the effects of the orientation (θ, ϕ) of the molecular axis and the position (X, Y) projected onto the surface are relatively small. As the molecule approaches the surface in a nearly perpendicular orientation ($\theta \cong 0^\circ$), only one atom of the diatomic molecule will interact with the surface. On the other hand, if the molecular axis is nearly parallel with the surface plane ($\theta \cong 90^\circ$), the two atoms of the diatomic molecule will simultaneously interact with the surface. As an example, the adiabatic potential energy surfaces on the copper surface obtained for parallel ($\theta = 90^\circ$) and perpendicular ($\theta = 0^\circ$) orientations are shown in Fig. 1.7. Here, (X, Y) is fixed on the bridge site (midpoint of two adjacent copper atoms), and ϕ is fixed at a specific angle . Assuming that the molecule moves along the direction perpendicular to the surface with its center-of-mass fixed on the bridge site, the resulting adiabatic potential energy surface depends on r and Z. If the hydrogen molecule is far from the surface ($Z \to \infty$), the equilibrium distance between the hydrogen atoms converges to $r = 1.40\,\text{a.u.}$, similar to the case of an isolated molecule. When the hydrogen molecule approaches the surface, there will be repulsion between closed electron shells due to the Pauli exclusion principle and the Coulomb repulsion between electrons, thereby increasing the potential energy. For the case of parallel orientation [see Fig. 1.7a], as the molecule approaches the covalent bond distance ($Z \cong 3.0\,\text{a.u.}$) ,

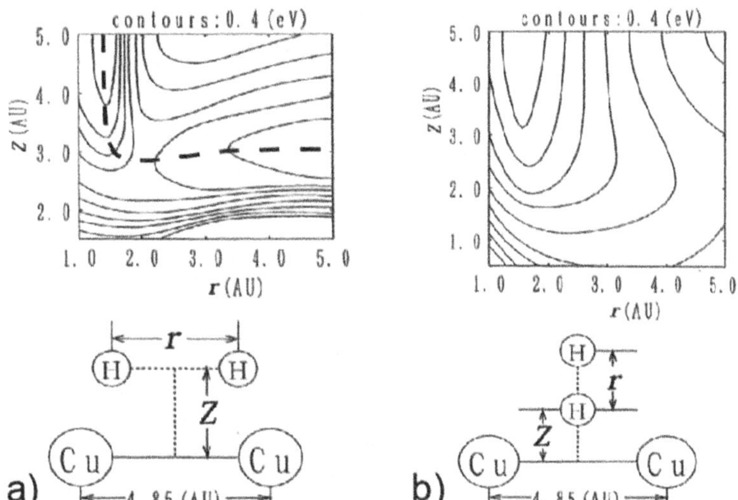

Fig. 1.7 Adiabatic potential energy surface expressed as a function of r and Z when the molecular axis is fixed **a** parallel and **b** perpendicular to the surface for hydrogen molecule motion on the copper surface. Space coordinates are displayed in atomic units. The contour interval is 0.4 eV. The broken line in (**a**) shows the reaction path determined from the figure. The figures below the contour plots show the definition of coordinates. Reprinted with permission from [21]. Copyright 2013 by Asakura Publishing

such increase in the potential energy could be suppressed by the increase in the internuclear distance. In addition, when the internuclear distance exceeds $r \cong 2.0$ a.u., the potential energy starts to decrease, and goes to the minimum at $Z = 3.0$ a.u., $r > 5.0$. Therefore, it is possible to connect the isolated molecular state and the dissociative adsorption state by drawing the path that traces the valley of the potential energy that crosses the saddle point. Conversely, when the molecular axis is oriented perpendicular to the surface [see Fig. 1.7b], the energy increases monotonically as the hydrogen molecule approaches the surface. This increase in the potential energy does not change even if the internuclear distance is increased. This means that the hydrogen molecule breaks the intramolecular bonds, and each hydrogen atom binds with surface atoms. The maximum point of the potential energy in the reaction path corresponds to a turning point where energy transfer occurs between translational motion (change of Z) and vibrational motion (change of r) of the molecule. In this example, the activation barrier is about 1.6 eV (with respect to the energy at $Z \to \infty$ shown in Fig. 1.7a).

In general, the incident molecule has finite translational and vibrational energies (also finite for quantum effects). Even if the total kinetic energies of the incident molecules are equal, the probability of adsorption differs if the ratio of translational energy to vibrational energy is different. Whether molecular vibrations promote or inhibit adsorption depends on the shape of the potential energy surface. Consider the case of a potential energy surface where the initial part of the reaction path is in the vacuum side. If molecular motion is treated classically, the trajectory when non-vibrating molecules are incident is shown in Fig. 1.8a. In this case, adsorption does not occur because the reaction path is against the barrier and subsequently, the molecule bends and is reflected back to the vacuum. On the other hand, when the

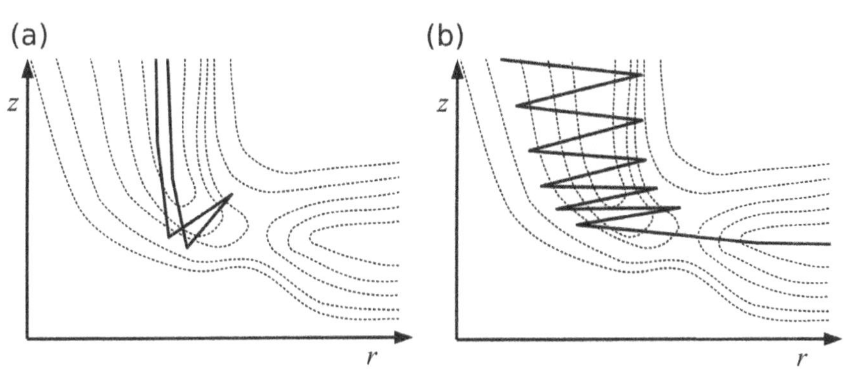

Fig. 1.8 A diagram showing the mechanism in which the vibration of the incident molecule assists the process of adsorption when the initial part of the reaction path is on the vacuum side of the activation barrier. The dashed curves show the contours of the adiabatic potential energy surface expressed as a function of r and Z. The thick solid lines show the classical dynamics trajectory of the incident molecule with the same kinetic energy as the activation barrier. **a** In the case where a non-vibrating molecule is incident. **b** When a vibrating molecule is incident. Reprinted with permission from [21]. Copyright 2013 by Asakura Publishing

molecule is vibrating, the path can follow the curve shown in Fig. 1.8b. This is called the *vibrationally assisted sticking* (VAS) effect. It can be understood that when the molecule is desorbed from the adsorbed state, molecular vibration is induced by following the reverse reaction path. This effect is called vibrational heating during desorption. On the contrary, for the case of a potential energy surface where the initial part of the reaction path is on the surface side, the inhibition of desorption by molecular vibration can be similarly analyzed.

The discussions so far have focused on two typical molecular axis orientations. However, in the actual adsorption/desorption processes, molecular axis rotation may occur to avoid the increase of potential energy. This phenomenon is called *steering*, and is one of the factors that assists the process of adsorption. The adiabatic potential energy surface expressed as a function of r and Z can be obtained for each inclination theta of the molecular axis as in Fig. 1.9, where a tentative reaction path can be obtained from the surface plots. If the potential energy on this tentative reaction path is expressed as a function of the one-dimensional coordinates s and θ defined along the path, the two-dimensional adiabatic potential energy surface shown in Fig. 1.9 can be obtained. Here, it can be assumed that the translational energy of the approaching molecule (for a given tilt of its axis) that is oriented perpendicular to the surfaceis lower than the activation barrier. However, its kinetic energy, including the rotational energy, is higher than the activation barrier. If steering does not occur, this molecule is reflected, due to the barrier, and can not be adsorbed (dotted line in Fig. 1.9). However, if steering occurs, the molecular axis may be reoriented (to lower the θ direction of potential energy) and the molecule will overcome the activation barrier (solid line in Fig. 1.9).

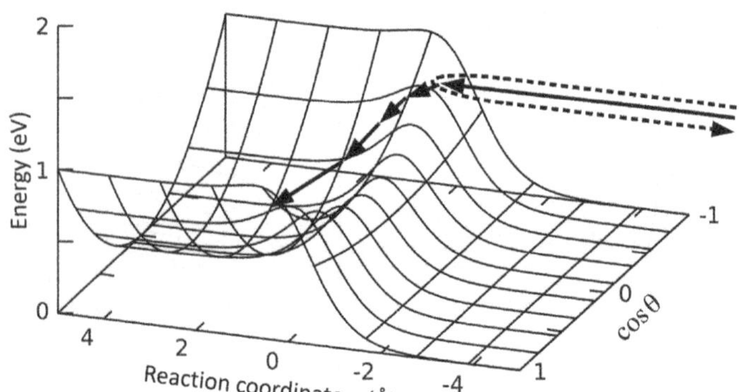

Fig. 1.9 Adiabatic potential energy surface as a function of the inclination of the molecular axis in hydrogen molecule motion on the copper surface. The vertical axis represents potential energy with the activation barrier at the origin. The variable s represents the coordinate on the reaction path where θ is fixed ($s \to \infty$ corresponds to $Z \to \infty$). The solid arrows schematically represent an example of the reaction path considering the steering effect in dissociative adsorption. The dashed arrows show an example where steering is not considered. The position of the arrow corresponds to the translational kinetic energy of the molecule. Adapted with permission from [21]. Copyright 2013 by Asakura Publishing

The likelihood of steering depends on the translational kinetic energy, vibrational energy, rotational energy and rotational axis of the incident molecule. When the rotational energy of the molecule is small, the momentum near the barrier becomes small and the direction of motion is easy to change, thereby making the steering process easy to occur. On the other hand, when the rotational energy of the molecule is sufficiently large, energy transfer between rotational and translational motions that support adsorption occurs. When the rotational energy $I\omega^2/2$ is large, the molecular rotation becomes a bound rotation close to the free rotation near the surface, and the angular momentum $I\omega$ is almost conserved. Here, I is the moment of inertia of the molecule and ω is the angular velocity. As the molecule approaches the surface, I increases as r increases, and the rotational energy decreases from $I\omega^2/2 = (I\omega)^2/2I$. The decrease in the rotational energy translates into translational motion. As a result, molecules will adsorb by acquiring translational energy that is higher than the activation barrier.

To summarize the effect of molecular rotation in dissociative adsorption, it is noted that the adsorption is assisted by steering in the small rotational energy regime, while there is energy transfer between the rotational and translational motions in the large energy regime. There will be no adsorption for all θ if the total kinetic energy is lower than the activation barrier. Here, when the area where the steering occurs and the area where the energy transfer occurs are clearly separated, the adsorption process with the assistance of the steering starts when the total kinetic energy exceeds the activation barrier for some θ. As the rotational energy increases, the steering effect disappears and the adsorption is inhibited. When the rotational energy further increases, adsorption begins to be promoted with the aid of energy transfer between the rotational and translational motions. For the desorption process, the desorption of molecules with small rotational energy is inhibited (rotational cooling), while the desorption of molecules with large rotational energy is promoted (rotational heating).

1.6 Atoms and Molecules Scattering

The beam scattering of ions, atoms, or molecules on surfaces is a phenomenon characterized by changes in the movement of these particles in the nanoscale region near the surface. It can be understood as a combination of elementary processes similar to adsorption and desorption. Beam scattering can be classified as inelastic collisions that involve changes in the translational energies of particles before and after the collisions Scattering that involves chemical reactions is specifically called **reactive scattering**. When translational energy is reduced during collision, the energy is converted into internal energy for the internal excitation due to the nuclear motion (such as the electron excitation energy of the surface or the colliding particle or vibration and rotation). Conversely, when the translational energy increases, the increase is at the expense of internal energy. The main internal degrees of freedom involved in energy conversion depend on the conditions of the incident particles (translation of particles, internal energy, etc.) and of the surface (temperature, etc.).

1.6.1 Ion Neutralization Scattering

A scattering that is accompanied by changes in the electronic state of the incident particle can occur if the surface or particle is charged or in an electronically excited state, and if the kinetic energy of the incident particle is large (10 eV or more). When the electron excitation is accompanied by electron transfer between the surface and the particle, neutralization and ionization of the particle can occur. An example of exchange of electron energy and nuclear translational kinetic energy in neutralization scattering of positive ions is shown in Fig. 1.10. When a positive ion approaches the surface, an overlap occurs between the valence electron wave function of the ion and the wave function protruding from the surface. This overlap enables electron transfer between the surface and the ions. When it is necessary for the electron system to acquire the energy δE in this electron transfer, the nuclear movement of the ion supplies such energy δE. As a result, ions incident to the surface and with kinetic energy E_{kin} are converted to neutral atoms and move away from the surface with a new kinetic energy $E_{\text{kin}} - \delta E$. Since electron motion is sufficiently faster than the ion motion, electrons may reciprocate between the surface and the ion for several times.

In particular, when the valence electron level E_{a} overlaps with the surface state or the band of the bulk, a resonance state is formed in which the spatial position and energy of the electron are mixed on the ion and the surface sides. As a result, electrons

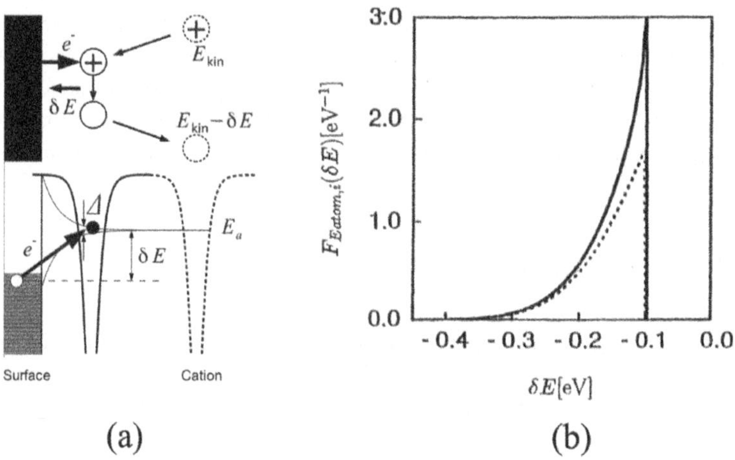

Fig. 1.10 **a** Schematic diagram of the neutralization mechanism in positive ion scattering, and **b** change in kinetic energy in the case of lithium ion with kinetic energy $E_{\text{kin}} = 100\,\text{eV}$. Scattering probability is expressed as a function of δE. The lower part of (**a**) shows the electronic structure and the electronic transitions that result in the neutralization. E_a is the valence electron level of the ion, and Δ is the life span generated by the resonance of the level with the electronic state of the surface. The broken line in (**b**) shows the value when the electron that jumps from the surface to the ion remains on the ion, and the solid line shows the value when the electron returns to the surface and then jumps back to the ion. Reprinted with permission from [21]. Copyright 2013 by Asakura Publishing

that have once entered the valence electron level will diffuse to the surface with high probability after a certain time (lifetime) has passed. At the same time, due to the uncertainty of time and energy, valence levels appear blurred with the lifetime width Δ. In the graph shown in Fig. 1.10, the loss of nuclear kinetic energy due to scattering is compared with and without considering the effect of this electron reciprocation mentioned. The difference between the solid and the broken lines in this graph shows the effect of electron reciprocation. The smaller the energy transfer amount δE, the greater the electron reciprocation effect due to the electron transition between states closer to the resonance condition. In order to understand the phenomena under these conditions properly, it is necessary to consider the electron reciprocation many times.

1.6.2 Rotational Excitation Scattering

Usually, the energies with respect to the molecule's internal degrees of freedom increase in the order of rotation, vibration, and electron motion. Consider the case where low-speed incident molecules with translational kinetic energy lower than the energy required for vibrational excitation collide with the surface and are scattered. Ignoring the changes in surface internal energy for simplicity, the translational kinetic energy of the incident molecule can only be converted into translational and rotational motion of the scattering molecule. The probability of rotational excitation depends on the incident conditions and the potential energy for molecular motion near the surface.

The scattering of NO molecules on Ag(111) is shown in Fig. 1.11 as an example. Assuming that the surface is flat and if the molecular axis at the time of incidence is perpendicular to the surface (θ with respect to the surface normal is 0° or 180°),

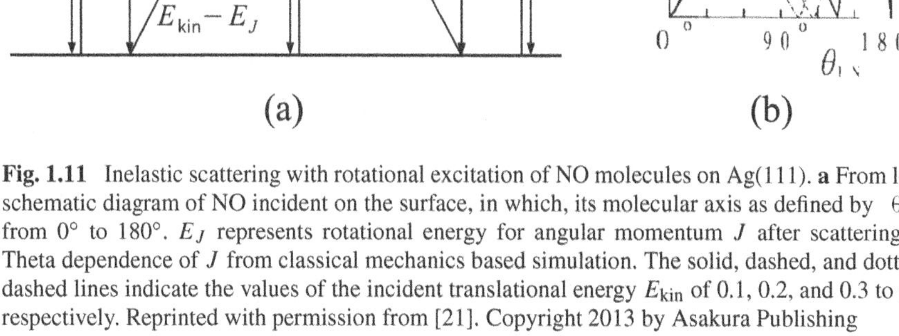

Fig. 1.11 Inelastic scattering with rotational excitation of NO molecules on Ag(111). **a** From left: schematic diagram of NO incident on the surface, in which, its molecular axis as defined by θ is from 0° to 180°. E_J represents rotational energy for angular momentum J after scattering. **b** Theta dependence of J from classical mechanics based simulation. The solid, dashed, and dotted-dashed lines indicate the values of the incident translational energy E_{kin} of 0.1, 0.2, and 0.3 to eV, respectively. Reprinted with permission from [21]. Copyright 2013 by Asakura Publishing

both N atoms and O atoms move vertically toward the surface. Thus, the molecule does not have rotational motion. When the molecular axis is tilted relative to the surface normal direction, the atom close to the surface experiences a force along the surface normal direction at the time of collision. On the other hand, the atom on the far side continues the inertial motion so as to maintain the intermolecular distance. As a result, a rotational motion of angular momentum J occurs in the molecule. The corresponding rotational energy E_J is generated from the incident translational kinetic energy E_{kin}, and the translational kinetic energy after scattering becomes $E_{\mathrm{kin}} - E_J$. When the molecular axis approaches the surface in parallel ($\theta = 90°$) orientation, the two atoms collide with the surface almost simultaneously. In such case, when a single diatomic molecule collides with a uniform surface, the atoms of the molecule experience the same force from the surface. The molecule, therefore, does not rotate, which characterizes an elastic scattering. As a result, the angular momentum of the scattering molecule has peaks at two theta values. In the case of heterodiatomic molecules, the elastic scattering angle deviates from $\theta = 90°$ because the force they received from the surface is different. Also, the deviation of the elastic scattering angle can be caused by the surface structure. In a classical mechanics based simulation shown in Fig. 1.11b, it is clear that the deviation of the elastic scattering angle θ from 90° is remarkable when the incident translational kinetic energy E_{kin} is small. The tendency is that lower incident translational energy E_{kin} has smaller momentum near the barrier. Thus, the potential due to the difference in the forces that acts on N and O atoms is susceptible to energy fluctuation.

Rotational excitation scattering can be understood from the perspective of steering. In addition, for the case of homogeneous diatomic molecules such as H_2, restrictions on the combination of the quantum state of rotation and nuclear spins are imposed by the symmetry of the wave function due to the requirement of quantum indeterminacy between homogeneous particles. At cryogenic temperatures below the rotational excitation energy, scattering with nuclear spin conversion occurs.

1.6.3 Vibrational Excitation Scattering

The hybridization of electronic states can be highly responsive to scattering. As an example, consider the scattering scenario with vibrational excitation of NO molecule on a high temperature Ag(111) shown in Fig. 1.12. NO molecule has the property to physically adsorb on Ag(111); but because its adsorption energy is small (~ 0.3 eV), the incident molecule hardly binds to the surface during the collision. Looking at the time-of-flight distribution of NO molecules before and after scattering, as shown in Fig. 1.12b, no correlation is found between the rotation/vibration and translational motion of the scattering molecules. Therefore, it can be seen that the kinetic energy of the scattering molecules is converted to surface vibration or electronic motion. Since the energy of surface vibration is smaller than molecular vibration by about an order of magnitude, the coupling with molecular vibration is expected to be mainly by electron movement.

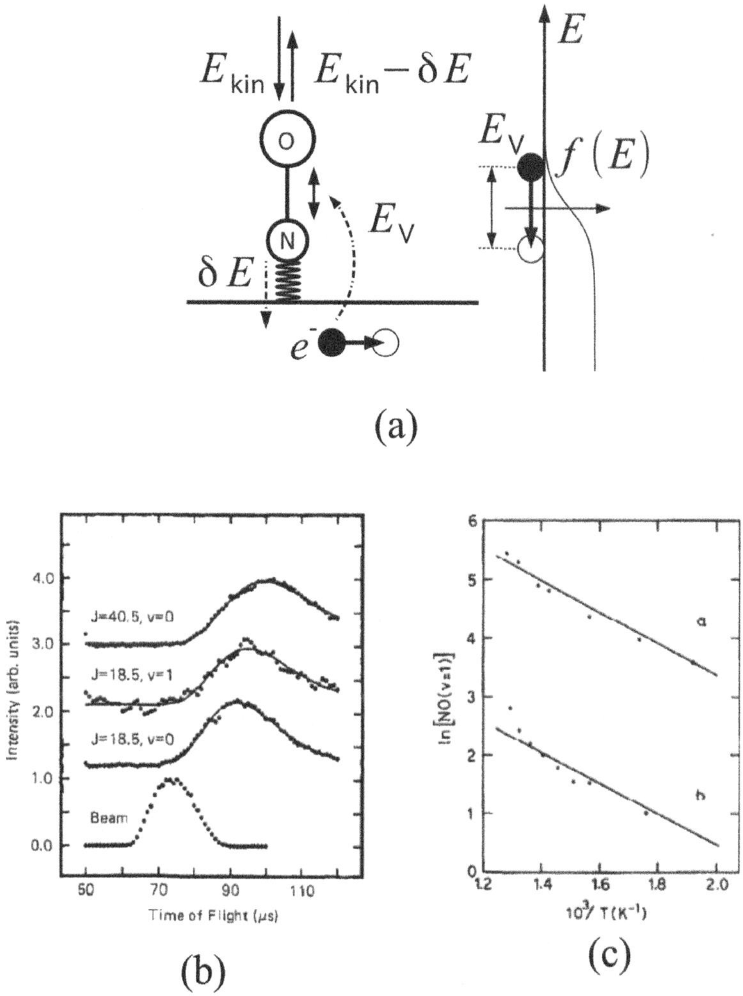

Fig. 1.12 Scattering with vibrational excitation of NO molecules on high temperature Ag(111). **a** A schematic diagram of a process in which an NO molecule moving with translational kinetic energy E_{kin} from a distance above the surface, absorbs an energy E_{υ} from surface electrons at the time of collision. It is then excited by vibration and is then scattered. Due to scattering, part of the translational kinetic energy δE is converted to surface vibration, etc. The right side is an energy diagram showing the process of combining electrons thermally excited according to the Fermi distribution $f(E)$ and supplying E_{υ}. The wavy lines between the molecule and the surface represent van der Waals interactions. **b** Time-of-flight distribution of NO molecules before and after scattering. The horizontal axis gives the flight time, and the vertical axis gives the beam intensity. The measurement results are for temperature of 760 K, incidence/measurement angle of 15°, and incident energy of $E_{\mathrm{kin}} = 0.94\,\mathrm{eV}$. J and υ indicate the rotational and vibrational quantum numbers of the scattering molecule, respectively. "Beam" is the data for the incident molecule. **c** The rate at which the NO molecule after scattering is excited from the ground vibrational state to the first vibrational excited state. The horizontal axis gives the surface temperature, and the vertical axis gives the logarithm of the ratio. **a** and **b** show the results for $E_{\mathrm{kin}} = 0.989, 0.084\,\mathrm{eV}$, respectively. The data points are from experiments while the solid lines are from the theoretical results. Reprinted with permission from [21]. Copyright 2013 by Asakura Publishing

Noting that the $2\pi^*$ orbital of NO molecule is close to the Fermi level of Ag(111), if the electrons and holes in the surface are thermally excited, movement to the $2\pi^*$ orbital can occur during collision. Because the $2\pi^*$ orbital is antibonding, when an electron occupies it, a force is generated in the direction that widens the distance between N and O. As a result, the hybridization of the electronic state between the surface and the molecule causes molecular vibrational excitation and coupling. That is, the thermally excited surface electrons jump into $2\pi^*$ orbitals and then combine with the surface holes to generate energy transfer E_V from the electron system to molecular vibration. On the other hand, the coupling between the electron system and the molecular translational motion is small because the bond between the molecule and the surface is not affected by the electron transfer since it is due to the van der Waals force. Figure 1.12c shows the rate at which the NO molecule after scattering is excited from the ground vibrational state to the first vibrational excited state as a function of the surface temperature. Although this ratio decreases logarithmically to the reciprocal of the surface temperature, this feature is consistent with theoretical calculations that take into account the coupling between the electron system and the molecular vibration [21]. Since this coupling can also bring about the reverse process of vibrational excitation, when the velocity of the incident molecule is slow (the translational kinetic energy E_{kin} is small) and the collision time is long, the energy is reversed back to the electron system and diffused, and the probability of vibrational excitation decreases.

When the time-of-flight distribution Fig. 1.12b is reviewed, it can be seen that the energy of the scattering molecule (corresponding to the reciprocal of the square root of the time of flight) is reduced compared to the incident molecule. This decrease in energy δE is considered to be due to energy transferred to surface vibration via van der Waals force.

1.6.4 Reactive Scattering

When atoms or molecules with kinetic energies of a few meV or eV strike the surface, both scattering and chemical reaction may occur. This phenomenon is generically called *reactive scattering*. The process where all atoms on the surface substrate remain on the surface after scattering is called *catalytic reaction*. On the other hand, a process in which atoms or adsorbed atoms are peeled off is called *stripping reaction*. Catalytic reaction is a basic process in technologies such as purification of exhaust gas, production of chemical substances, and batteries (the oxidation-reduction reaction at the electrode produces an electromotive force). Stripping reaction is a basic process in technologies such as organic synthesis from inorganic substances (such as formic acid synthesis from hydrogen and carbon dioxide) and etching (surface processing using corrosion). Generally, in reactive scattering, much of the energy generated by a chemical reaction is transferred to the kinetic energy of the scattering molecule. Thus, the scattering molecule has a kinetic energy that is higher than that at the time of incidence (i.e., the temperature rises).

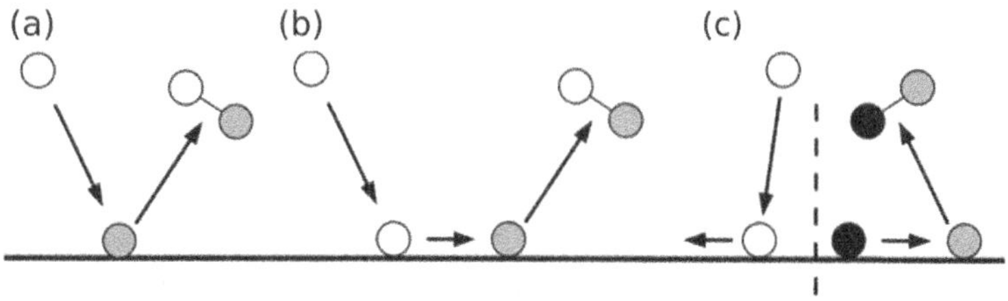

Fig. 1.13 Mechanism of stripping reaction. Reprinted with permission from [21]. Copyright 2013 by Asakura Publishing

Catalytic and stripping reactions differ only in the final state of the surface atoms, but the theoretical analysis is essentially similar for both cases. Since the stripping reaction involves changes in the number of atoms on the surface, macroscopically, it is a non-equilibrium phenomenon in which atoms diffuse from the surface to a highly fluid phase (gas phase in the case of a solid-gas phase interface). Several mechanisms have been proposed microscopically. The mechanism by which atoms and molecules approach the surface and peel off the surface atoms directly, leaving the surface without being adsorbed is called the *Eley-Rideal mechanism* [Fig. 1.13a]. The mechanism by which atoms and molecules that move to the surface are adsorbed on the surface and diffuse with high energy, and then the surface atom is stripped off is called the *hot atom mechanism* [Fig. 1.13b]. The mechanism by which atoms and molecules that approach the surface temporarily adsorb and reach quasi-thermal equilibrium and then peel off the surface atoms before finally leaving the surface is called the Langmuir-Hinshelwood mechanism [Fig. 1.13c]. Analysis based on adiabatic potential energy is effective regardless of which mechanism is involved.

References

1. H. Kasai, H. Akai, H. Yoshida (eds.), *Keisanki Materials Design Nyumon* (Osaka University Press, Osaka, 2005). (in Japanese)
2. H. Kasai, M. Tsuda, *Computational Materials Design Case Study 1: Intelligent/Directed Materials Design for Polymer Electrolyte Fuel Cells and Hydrogen Storage Applications* (Osaka University Press: Osaka, 2008) (in Japanese)
3. H. Kasai, H. Kishi, *Computational Materials Design Case Study 2: Intelligent/Directed Materials Design for Resistance Random Access Memory Device* (Osaka University Press, Osaka, 2012). (in Japanese)
4. H. Kasai, M. Sakaue, *Surface and Interface Physics* (Asakura Shoten, Tokyo, 2013). (in Japanese)
5. H. Kasai, M.C.S. Escaño (eds.), *Physics of Surface, Interface and Cluster Catalysis Bristol* (IOP publishing, UK, 2016)
6. R. Kishida, H. Kasai, *Computational Materials Design Case Study 3: Biosynthesis of Melanin Pigment* (Osaka University Press, Osaka, 2019). (in Japanese)
7. P. Hohenberg, W. Kohn, Phys. Rev. B **136**, 864 (1964)
8. L.H. Thomas, Proc. Cambridge Phil. Soc. **33**, 542 (1927)
9. E. Fermi, Rend. Lincci **6**, 602 (1927)

10. J. Perdew, K. Burke, M. Ernzerhof, Phys. Rev. Lett. **77**, 3865 (1996)
11. J. Perdew, K. Burke, M. Ernzerhof **78**, 1396 (1997)
12. J.P. Perdew, Y. Wang, Phys. Rev. B **33**, 880 (1986)
13. A.D. Becke, Phys. Rev. A **38**, 3098 (1988)
14. C. Lee, W. Yang, R.G. Parr, Phys. Rev. B **37**, 785 (1988)
15. K.W. Kolasinski, *Surface Science: Foundations of Catalysis and Nanoscience*, 2nd edn. (Wiley, 2008)
16. R.L. Arevalo, S.M. Aspera, M.C.S. Escano, H. Kasai, ACS Omega **2**, 1295–1301 (2017)
17. R.L. Arevalo, M.C. Escano, H. Kasai, ACS Catalysis **3**, 3031–3040 (2013)
18. P. He, X. Wang, Y. Liu, L. Yi, X. Liu, Int. J. Hydrogen Energy **37**,1254–1262 (2012)
19. V.R. Stamenkovic et al., Science **315**, 493–497 (2007)
20. M.C.S. Escano, H. Nakanishi, H. Kasai, J. Phys. Chem. A **113**, 14302–14307 (2009)
21. H. Kasai, M. Sakaue, *Surface and Interface Physics* (Asakura Publishing, Shinjuku, Tokyo, Japan, 2013)

Chapter 2
Behavior of Hydrogen and Hydrogen-Containing Molecules on Metal Surfaces

Abstract As the lightest element, hydrogen is typically used as a test particle to establish the fundamental concepts in surface science and to evaluate surface reactivity and relevant surface processes. Aside from this, hydrogen attracts enormous interests for industrial and technological applications due to its potential use as the key energy carrier in the future. If realized, it will provide an environmental friendly alternative to conventional fossil fuels. This chapter aims to discuss the interaction of hydrogen and other compounds such as methane and water with transition metal surfaces. Also, induced surface reconstructions and induced surface segregations in bimetallic systems will be tackled. Such processes are usually taking place on surfaces in the presence of adsorbates. As much as possible, the role of metal surfaces in hydrogen-related applications will be evaluated based on their reactivities and properties that could possibly bring important ideas to further improve existing technologies that will aid in the realization of hydrogen-based technologies.

2.1 Hydrogen Atom Adsorption and Hydrogen Molecule Dissociation

The interaction of hydrogen with metal surfaces has been widely reported in the literature [1–9]. For most transition metal surfaces, hydrogen molecule (H_2) readily dissociates and the dissociated H atoms are strongly adsorbed on the surfaces. The activated dissociation of H_2 is known to take place on coinage metals such as Cu, Ag, and Au, which is attributed to their unique electronic properties compared to other transition metals. Because these coinage metals have fully occupied *d*-bands, it is difficult for charge to transfer from the metal to the H_2 antibonding orbital. On the other hand, for transition metals with partially filled *d*-bands such as Pd or Pt, the electrons of the metal surface can readily populate the H_2 antibonding orbital due to the presence of *d*-states in the Fermi region. Such population of H_2 antibonding orbital facilitates the H–H bond dissociation. Moreover, the good reactivity of the transition metals with partially filled *d*-band is also evident on the strength of adsorption of the dissociated H fragments. In particular, the adsorption of H atom is stronger on

© Springer Nature Singapore Pte Ltd. 2020

H. Kasai et al., *Hydrogen and Hydrogen-Containing Molecules on Metal Surfaces*, Springer Series in Surface Sciences 71,
https://doi.org/10.1007/978-981-15-6994-4_2

Pd and Pt surfaces than on Cu, Ag and Au. The *d*-bands of Pt(111) and Cu(111) as representative metal surfaces are shown in Fig. 2.1 for comparison.

The electronic structure of an H_2 molecule as it approaches a metal surface is shown in Fig. 2.2a. The corresponding contour plot of the potential energy surface

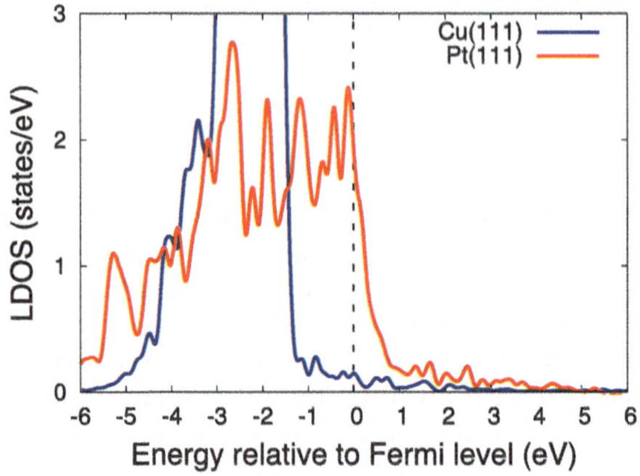

Fig. 2.1 Local density of states (*d*-band) of Cu and Pt atoms on Cu(111) and Pt(111) surfaces, respectively. It shows that the states of Cu are lying below the Fermi level while there are states in the Fermi region for the case of Pt

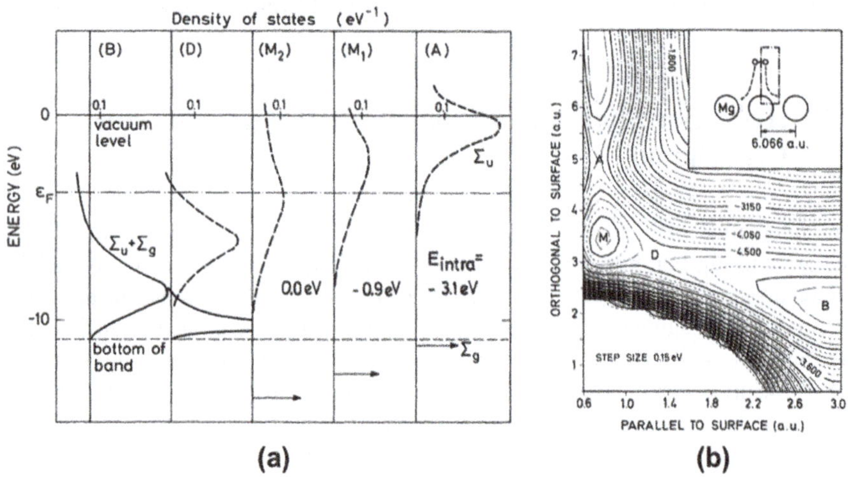

Fig. 2.2 **a** Electronic structure of H_2 orbitals as the molecule approaches a metal surface. The panels (left to right) show the electronic structures of H_2 moving closer to the surface. **b** Contour plot of the potential energy surface of an H_2 that approaches the surface. Molecular adsorption state is observed at M and proceeds to the dissociative adsorption state at B, with points A and D as the transition states. Reprinted with permission from [10]. Copyright 1981 by the American Physical Society

(PES) is shown in Fig. 2.2b. As H_2 approaches the surface, the antibonding orbital (Σ_u) of the molecule broadens and shifts below the Fermi level, indicating that it is populated during the adsorption process. This results to the weakening of the H–H bond and leads to the dissociative adsorption process. During such process, the H-metal interaction becomes stronger while the H–H interaction is weakened. As shown in the contour plot of the PES, H_2 is molecularly adsorbed on the metal surface, with a transition state for molecular adsorption defined at point A. It then dissociatively adsorbs on the surface defined by point B, with a corresponding transition state at point D.

For most fcc metal surfaces with (111) and (100) facets, an H atom favors to adsorb on the three-fold hollow and four-fold hollow sites, respectively. However, deviation from these preferred adsorption sites may happen for metals with large lattice constants. For example the adsorption of H on Au(100) is more favored on the bridge site than on the hollow site [11]. Note that Au has larger lattice constant than other fcc metals such as Pt, Pd, Cu, and Ni. It can be deduced that for the case of Au surface, the bond length between the surface Au atoms and H will be optimized if H sits on the bridge site rather than on the four-fold hollow site.

Aside from the kinds of metals, the adsorption energy of an H atom also depends on the surface facets. Adsorption energies are stronger on more open surfaces such as fcc(110), compared to more closed-packed surfaces like the fcc(111). For example, the adsorption energies of an H atom on Pd(110) and Pd(111) surfaces are −0.50 eV and −0.45 eV, respectively [12]. Also, this observation can be understood from the different characteristics of the *d*-bands of the surface atoms. Among the low-indexed surfaces of the fcc crystal, the (110) facet has the most open structure. As such, the surface atoms on (110) facet are less influenced by other surface atoms due to the larger interatomic distances as compared to the closely arranged atoms on the (111) and (100). Because of this weaker surface atom to atom interaction, the *d*-band tends to be more localized and shifts closer to the Fermi region compared to the *d*-band of the more closed-packed surface atoms. This upshift of the *d*-band towards the Fermi region is a consequence of the necessity to retain the number of electrons that is implicitly defined in the *d*-band. With such property, it can be depicted that more open surfaces have *d*-band centers that are closer to the Fermi level. The *d-band center* is a representative value of the weighted average of the *d*-band of a metal surface. It is typically used as a descriptor or an indicator of the reactivity of metal surfaces toward an adsorbate [13]. More reactive metal surfaces have *d*-band centers closer to the Fermi level. It can be calculated once the density of states is obtained, and is defined by the equation

$$\mathcal{E}_d = \frac{\int_{-\infty}^{\infty} n_d(\mathcal{E})\mathcal{E}d\mathcal{E}}{\int_{-\infty}^{\infty} n_d(\mathcal{E})d\mathcal{E}}. \tag{2.1}$$

Here, n_d is the density of states at particular energy state $\mathcal{E}$ [14].

Studying the interaction of gas molecules on stepped metal surfaces is important because it resembles realistic systems for industrially employed metal catalysts.

Surface corrugation and the availability of kinks, steps, terrace and edges provide different reactivities with respect to the symmetry sites offered by flat surfaces. Hence, understanding the interaction of molecules on stepped metal surfaces is essential in catalytic reactions that find indispensable applications in chemical industries.

Several studies have shown an enhancedadsorption of H atom on stepped metal surfaces over the flat metal surfaces. Aside from the changes in the adsorption energies, the preferred adsorption sites of H are also modified on stepped surfaces. For example, in the case of Pt(533) surface shown in Fig. 2.3a, the adsorption of H is most favored on the bridge (B1) site of the step edge with a corresponding adsorption energy of −0.41 eV [15]. Here, the adsorption free energy of H is obtained by considering the zero-point corrections to the total energy. (A more negative adsorption energy indicates stronger adsorption on the surface.) On flat Pt(111), H favorably adsorbs on the fcc hollow site with an adsorption energy of −0.29 eV. While H can adsorb on the bridge site of Pt(111), it might not adsorb on the bridge site in the terrace of Pt(533). This is due to the presence of energy gradient that allows the movement of H atom from the bridge site towards the neighboring site, i.e. hollow site, which provides stronger adsorption for H. There is also an enhanced adsorption energies on on-top site at lower step edge of the (533) surface (OT4 site in Fig. 2.3a) compared to the corresponding adsorption energy of H at on-top site of (111) or at the terrace of the (533) surface. Clearly, the surface atoms of the step-edge region of the (533) surface provide larger H adsorption energies than those of flat Pt surface.

While the enhancement of H adsorption energies on the step edge of Pt(533) has been observed with respect to its adsorption on Pt(111), the adsorption of H on the step edge of Pt(553) did not exhibit the same trend (Fig. 2.3b). Stronger adsorption is observed on the step edge than on the terrace of the (553) surface. However, the adsorption energies are comparable with the H adsorption energies on Pt(111). Adsorption energy is largest on the FCC1 site, which is located at the step edge region. In such case, adsorption energy is found to be −0.28 eV, which is comparable to the adsorption energy on FCC site of Pt(111). Also, the adsorption of H on on-top sites is not enhanced with reference to on-top site adsorption on Pt(111), unlike the case of H adsorption on Pt(533). These results show that the enhancement of H adsorption

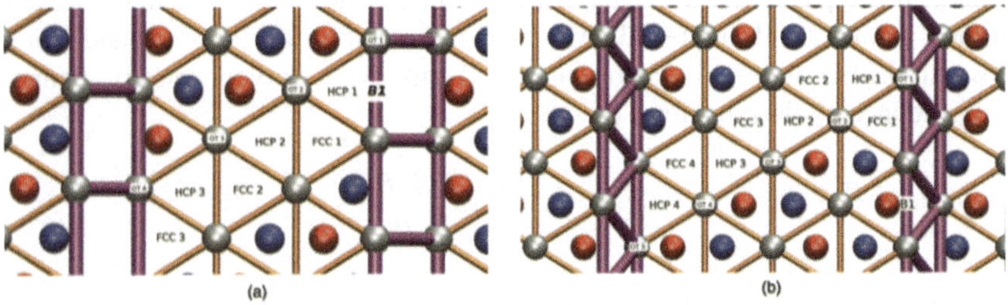

Fig. 2.3 Adsorption sites of H on **a** Pt(533) and **b** Pt(553) stepped surfaces. The atoms in the topmost, second and third layers are represented by silver, blue, and red spheres, respectively. Reprinted with permission from [15]. Copyright 2014 by the AIP Publishing LLC

energy on stepped surface over the flat surface is not a general feature or behavior of stepped surfaces. It demonstrates however, that the edge sites are more reactive than the terrace sites as observed on both Pt(553) and Pt(533) surfaces.

Studies on the dissociation of H_2 on metals found lower activation barrier of dissociation on the stepped surfaces compared to flat surfaces. For instance, the dissociation of H_2 on Cu(211) requires 0.41 eV of activation barrier, compared to the 0.46 eV barrier on Cu(111) surface. The fcc(211) surface is a stepped surface formed by two (111) terraces that are separated by a monatomic step. In such case, the step edge is involved in the dissociation process, which is responsible for the lowering of the activation barrier [16]. On the other hand, the dissociation of H_2 on stepped Pt(211) surface is found to be non-activated if the center of mass is directed at the on-top site of Pt residing at the step [17]. On the terrace region of the surface, an activated dissociation of H_2 occurs. The chemisorption wells that exist above the lower edge of the steps could serve as an attractive channel for approaching molecules. This explains the lower dissociation barrier of H_2 on the stepped region.

There is also an observed trend in the reactivities of stepped metal surfaces with its *d*-band center. As mentioned earlier, there is a lower dissociation barrier on stepped Pt (with partially filled *d*-band) than on stepped Cu (completely filled *d*-band). Figure 2.4 shows the local density of states (LDOS) projected on the *d*-states of step atoms on Cu(410) and Pt(410) surfaces. Compared to the LDOS of the atoms in the terrace, the shift of the *d*-bands of the atoms in the steps to higher energies can be observed. This is due to the lower coordination of the atoms in the steps. Consequently, it is expected that the d-band center will be more positive compared to the *d*-band centers of the flat metal surfaces. The *d*-band centers of the atoms on the terrace and on the steps of Cu(410) are −2.24 eV and −2.11 eV, respectively. The corresponding values for Pt atoms on Pt(410) are −2.17 and −2.00 eV respectively. The observed reactivity of the step atoms can therefore be understood from the values of *d*-band centers.

Aside from steps, kink regions on surfaces can also provide lower dissociation barrier for H_2 with reference to its dissociation on flat surfaces. Kink regions are

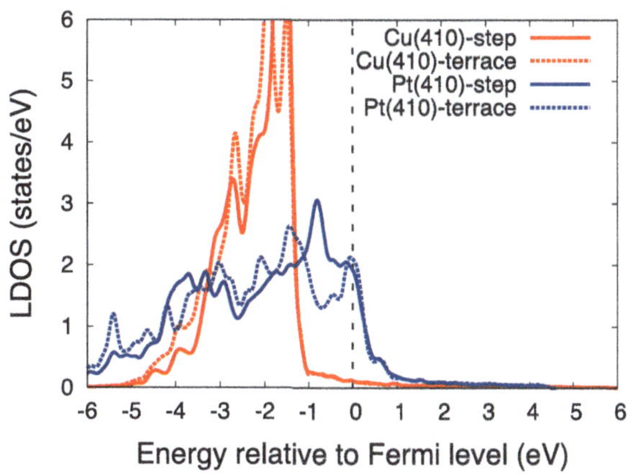

Fig. 2.4 Local density of states (*d*-states) of Cu and Pt atoms on Cu(410) and Pt(410) surfaces. The LDOS of atoms in the steps are shifted to the right compared to the LDOS profiles of Cu and Pt atoms in the terrace region of the surfaces

usually regarded as defect sites on surfaces, which are usually present in stepped or vicinal surfaces. The dissociation paths of H_2 on these defective and stepped surfaces are shown in Fig. 2.5 [18]. Here, the center-of-mass of the dissociating H_2 is marked by B (bridge), TB (between top and bridge), or T (top). Results have shown that the dissociation barrier is lowered by ~0.30 eV when the dissociation process happens on the kink site (Fig. 2.5d), with respect to its dissociation on Cu(111) (Fig. 2.5a). In this model, the kink sites are introduced in Cu(211). Similar lowering of the dissociation barrier is observed when the path of H_2 is directed on the vacancy defect site of Cu(111). More importantly, with respect to the dissociation process on Cu(111), lower activation barriers are obtained on the stepped regions of Cu(211) and Cu(221) surfaces. However, the activation barriers are observed to be path dependent, indicating the influence of local surface structure on the dissociation process of H_2. Such observation is evident on the slightly lower barriers for dissociation on TB sites than on B sites of Cu(211) and (221) as shown in Fig. 2.5b, c. The dependence of the activation barrier of H_2 on its dissociation path is also found in Cu(111). Lower

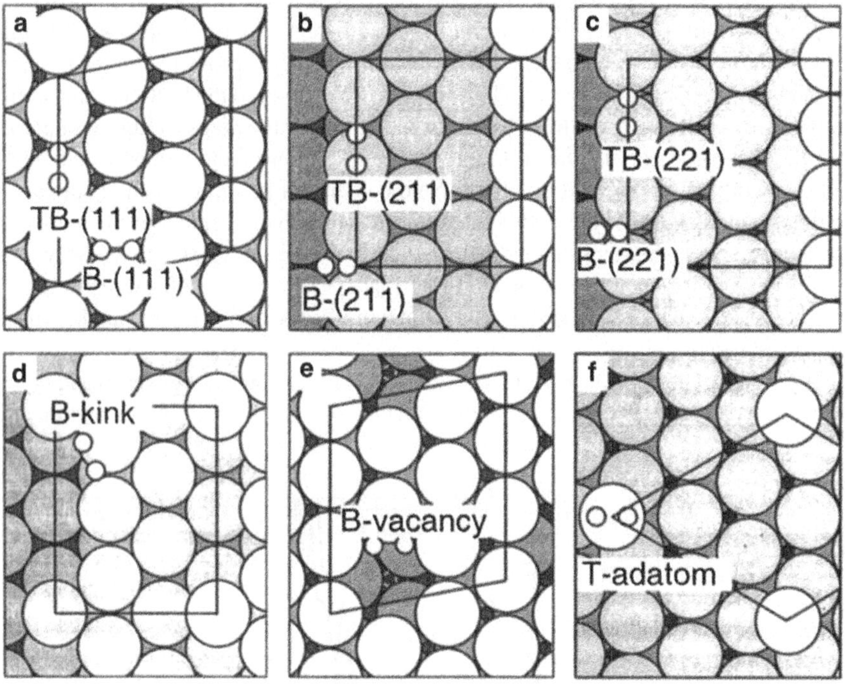

Fig. 2.5 Top views of **a** Cu(111); **b** Cu(211); **c** Cu(221); **d** kink at Cu(211); **e** vacancy at Cu(111); **f** adatom at Cu(111). Large and small circles represent Cu and H atoms, respectively. The center-of-mass of H_2 relative to its position on the surface is defined as B (bridge), TB (between top and bridge), or T (top). Reprinted with permission from [18]. Copyright 2002 by the American Physical Society

activation barrier exists on the B site with an activation barrier of 0.55 eV than on the TB site with an activation barrier of 0.89 eV.

These observed variations in the activation barriers are clarified by analyzing the electronic properties of the surface Cu atoms on these different surfaces. The change in the 4*sp* electronic states of Cu is identified to be a more important factor than the changes observed in its 3*d* states. Note that the comparison is among the Cu atoms that are interacting with the incoming H_2. Specifically, the overlap of the bonding and anti-bonding states of H_2 molecule and Cu atoms in the bridge or top sites is measured. The variations in the calculated barriers on the bridge (B) sites in Fig. 2.5a, d are described by the smaller overlap in the bonding states on the B-kink. Such smaller overlap implies lesser repulsion and lower stability. Thus, lower energy is necessary to allow the participation of the surface electrons in the reaction, which leads to lower activation barrier on B-kink than on B site. On the other hand, the lower barrier on B-kink than on between top and bridge (TB) site of Cu(211) is more dependent on the overlap of the antibonding states. There is a larger overlap in the B-kink for such state, which facilitates an easier dissociation of the molecule. This shows that the reactivities of the steps and kinks are highly determined by the electronic properties of the surface and the impinging H_2 molecule.

While most studies suggest the reactive nature of the step atoms to be responsible for the lowering of activation barriers in the dissociation of H_2, there are recent studies which argue that the terrace or flat regions are more reactive than the steps toward H_2 dissociation. In a combined molecular beam experiments and quasi-classical trajectory (QCT) simulations obtained from potential energy surface (PES) calculations, it was reported that the Cu(111) surface is more reactive than the Cu(211) [19]. Despite the fact that the trend in the *d*-band centers of the surface atoms at the stepped and flat surfaces were found to be consistent with previous works (more positive *d*-band center for atoms in the step), the calculated barrier is lower on the flat Cu(111) surface than on the stepped Cu(211) surface. Experimentally, this was verified via the higher sticking probability of D_2 on Cu(111), which implies that the surface has higher reactivity toward D_2 dissociation (Fig. 2.6). Such phenomenon was attributed to "geometric effects", where different dissociation paths for H_2 on Cu(211) and Cu(111) can exist. The experimental findings are likewise verified by the calculated activation barriers. The activation barrier for H_2 dissociation on the steps of Cu(211) with its center-of-mass near the top site is 0.678 eV. On the other hand, 0.636 eV activation barrier accompanies H_2 dissociation on Cu(111) with the center-of-mass directly on the bridge site. The *d*-band center theory still works if the comparison for dissociation is made for the same dissociation centers (e.g. near top site of (211) with barrier of 0.678 eV versus the near top site of (111) with barrier of 0.837 eV). However, the measurement of reactivity requires the consideration of all possible dissociation pathways and dynamics on the surfaces.

The top view of a (1×1)-Cu(211) surface is shown in Fig. 2.7a, while the corresponding energy heat map for activation barriers of H_2 dissociation is shown in Fig. 2.7b. It can be noted that low barriers exist on the Cu(111)-like b_2 and on t_2b sites at the middle region, as well as on the regions at the steps (t_1 and b_1 sites). The t and b denote top and brige, respectively. High barriers exist between these sites

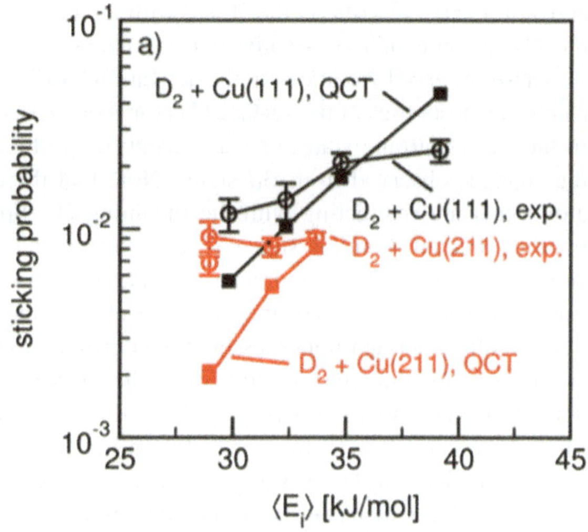

Fig. 2.6 Experimental and calculated sticking probabilities for D_2 on Cu(111) (black) and D_2 on Cu(211) (red) as a function of incidence energy (E_i). The error bars indicate standard deviations from 10 measurements. Reprinted with permission from [19]. Copyright 2017 by the American Chemical Society

(bottom region), although there is a localized region identified at t_2 site. In these calculations, the activation energies at t_2b, t_1, t_2, b_1 and b_2 sites are 0.678, 0.696, 0.721, 0.707, and 0.671, respectively, in units of eV. These values demonstrate the slightly lower barriers at the middle region as compared to the dissociation on the steps. These findings likewise complement the higher sticking probability and lower dissociation barriers obtained on flat Cu(111) than on stepped Cu(211).

These results illustrate the complex reactivity of stepped metal surfaces toward the dissociative adsorption of an H_2 molecule. While the H and H_2 have relatively simple geometric and electronic properties, the results from the presented scientific investigations demand more intensive studies to clarify the reactivity of stepped metal surfaces. With the aid of more advanced experimental equipment and more powerful computational tools, it is expected that new findings on metal—hydrogen system will be unraveled in the near future.

2.2 Water Molecule Adsorption

The adsorption of water molecules on surfaces is a key elementary step in many catalytic reactions and of profound importance in many physical, chemical, and biological phenomena. Water as a solvent is typically present in reactions that are carried out in laboratories, and is considered to be the most economical source of hydrogen for energy related applications. Because metals are often used as catalysts in chemical industries for various applications, it is imperative to elucidate the mechanism of water interaction on transition metals.

The interaction of water with surfaces is more complicated compared to the earlier discussed interaction of diatomic molecules on surfaces. Aside from having three atoms, water as a free molecule has a bent structure. This complicates the simulation

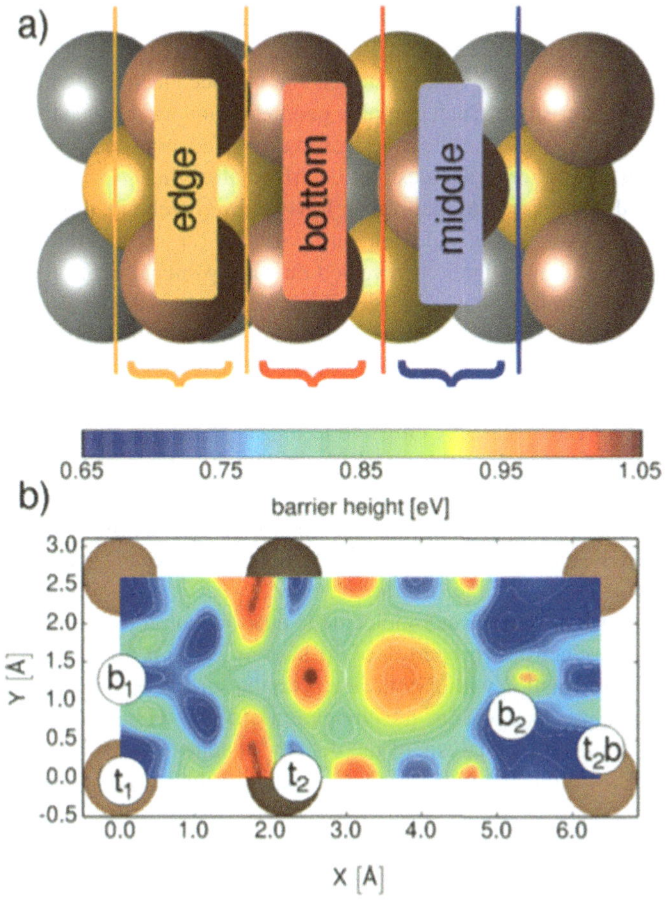

Fig. 2.7 **a** Top view on the (1 × 1) unit cell of Cu(211), and **b** optimized barrier heights as a function of the impact site according to the 6D PES (t and b denote top and brige, respectively). Reprinted with permission from [19]. Copyright 2017 by the American Chemical Society

of its adsorption due to the necessary considerations of the symmetry and degrees of freedom of the molecule. Unlike the case of a linear molecule, the orientation of water should be defined with reference to its bent structure and of its component atoms. For instance, a linear molecule (e.g. CO_2) could be positioned parallel or perpendicular to the surface plane, and its azimuthal and polar angles with respect to the surface could be varied. The number of possible configurations will be less than the case of a molecule with bent structure. For the case of H_2O, the position of the bent (or the direction of the "V-shaped" structure of H_2O) should also be considered in addition to its orientation (parallel or perpendicular) relative to the surface.

The bent geometric structure of H_2O can be attributed to the distribution of its electrons. Two pairs of valence electrons are responsible to the formation of two

O–H covalent bonds, while two lone pairs are present in the molecule that are distributed around the oxygen atom. These lone pairs make the O region negatively charged. The nucleus of the O atom attracts the valence electrons, which makes the H atoms partially positive. Such complex geometric and electronic structures of the isolated H_2O molecule make the investigation of its adsorption on surfaces and its co-adsorption with other molecules particularly challenging.

Theoretical and computational research works on the elucidation of the adsorption structures of water molecules on surfaces were reported in the literature. On Pt(111), DFT-based investigations revealed the on-top site adsorption of H_2O (with the position of O atom as the reference). Figure 2.8 shows the adsorption configuration of a water monomer on the surface [20]. The molecule is almost parallel to the surface plane with a tilt angle of 8°. In such configuration, the adsorption energy of the molecule is −0.30 eV and could be classified as either weakly chemisorbed or physisorbed on the surface. Compared to its gas-phase structure, the structure of H_2O is not significantly modified [21]. The O–H bond of adsorbed water is almost similar to the 0.97 Å O–H bond in its gaseous phase. The HOH angle on the other hand has increased only by 2°. In addition, the metal atom immediately under the H_2O molecule is displaced along the direction of the surface normal with respect to the other surface atoms on the surface.

The preference of single water molecule to adsorb at on-top site can be explained by analyzing the electronic properties of the system. Water is found to be an electron donor upon its interaction with the surface, with the surface atoms acting as electron acceptors. Such process explains the elongation of the O–H bond and the widening of the HOH angle [22]. Analysis of the density of states clarified the origin of the preference of the molecule to be oriented parallel to the surface plane. Figure 2.9a

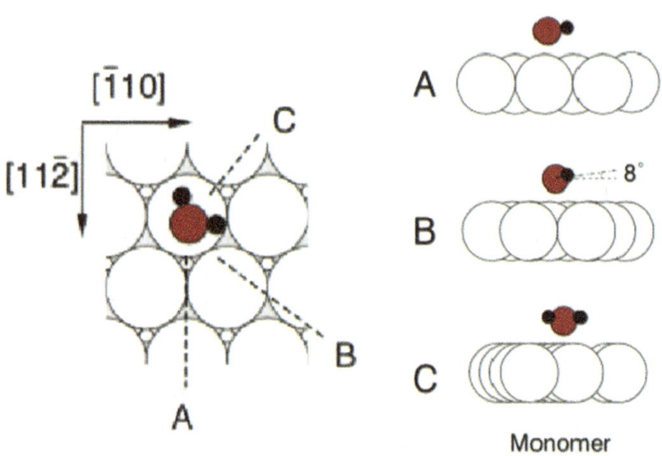

Fig. 2.8 The adsorption orientations and configurations of a water molecule on Pt(111). A, B and C planes are defined such that the side views of the structures could be easily identified. The Pt, O and H atoms are represented by white, red and black spheres, respectively. Reprinted with permission from [20]. Copyright 2010 by Elsevier

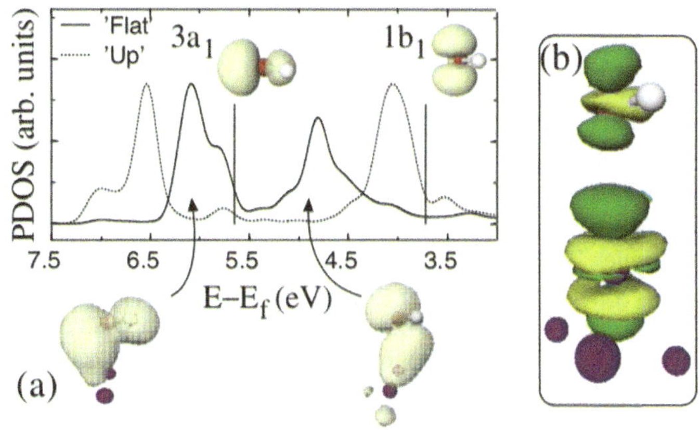

Fig. 2.9 **a** Partial density of states (*p* states) of the O atom in H_2O, which demonstrate the $3a_1$ and $1b_1$ states of the adsorbed molecule. 'Flat' (solid line) and 'Up' (dotted line) signify the PDOS of the molecule that is adsorbed in parallel and perpendicular orientations. **b** Charge density difference of adsorbed H_2O on Pt(111). Dark (light) regions correspond to a density decrease (increase) of 3.6×10^{-2} eA^{-3}. Reprinted with permission from [21]. Copyright 2003 by the American Physical Society

shows the partial density of states (*p* states) of the O atom in H_2O, indicating the $3a_1$ and $1b_1$ states of the adsorbed molecule [21]. 'Flat' (solid line) and 'Up' (dotted line) signify the PDOS of the molecule that is adsorbed in parallel and perpendicular orientations, respectively. The plot shows that the $3a_1$ of the Up configuration is positioned at lower energies than the Flat configuration, which suggests that the $3a_1$ state is more stable in the Up than in the Flat configuration. However, the $1b_1$ state of the Flat structure is at lower energies than the corresponding state for the Up region. Since the $1b_1$ state is closer to the Fermi region, such state could be considered as the one that maximizes the interaction with the surface; hence, the preferred parallel orientation of the adsorbed water. The charge difference distribution shown in Fig. 2.9b shows the $1b_1$-like state of the adsorbed molecule on the surface, which is consistent to the analysis of the PDOS presented in Fig. 2.9a.

The adsorption of water was found to be stronger on transition metals with partially occupied *d*-band than on the coinage metals with fully occupied *d*-band. This is consistent with the earlier discussion on the trend in adsorption energies of molecules with the electron occupancy of the *d*-band. The reported trend in the magnitude of adsorption energies of water is Au < Ag < Cu < Pd < Pt < Ru < Rh [21]. In particular, the adsorption energy of water on Rh(111) and Au(111) are −0.42 eV and −0.13 eV, respectively. These data show that despite the complexity in the structure of H_2O as compared to H_2, it obeys the predictions of the *d*-band theory. The optimum bond length between the O atom of H_2O and the nearest metal atom (M) reflects the strength of H_2O adsorption. For example, the O–M distances on Rh(111) and Au(111) surfaces are 2.31 Å and 3.02 Å, respectively. This demonstrates a closer distance of H_2O on the surface for the case of a stronger adsorption. Notably, regardless of the type of the metal surface, H_2O generally prefers the parallel orientation with respect to the surface plane.

Water clusters are formed via the hydrogen bonds between the molecules mediated by the lone pairs in the O atom and the partially positive charged H atoms of the molecule. In the cluster, the hydrogen atom of one molecule interacts with the O atom of another molecule. In the simplest case of water cluster called the water dimer, one molecule acts as a hydrogen bond donor, while the other molecule is the acceptor molecule. The adsorption configuration of a dimer on Pt(111) surface is shown in Fig. 2.10. The figure shows the interaction between the molecule via the hydrogen bond as well as the interaction between the molecules and the surface [20, 22]. Specifically, the O atom of the donor molecule is closer to the Pt surface than that of the acceptor molecule. In addition, the donor molecule has geometric properties similar to that of the single molecule adsorbed on the surface. With respect to the surface plane, the donor molecule is tilted upward just like the monomer case. On the other hand, the acceptor molecule is tilted downward. In general, parameters such as the bond angles and the O–H interatomic distances have small variations with respect

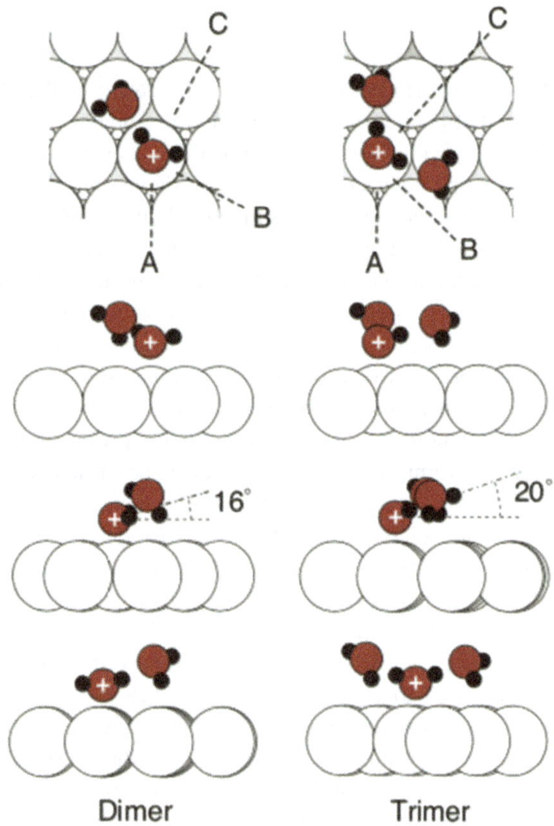

Fig. 2.10 Adsorption orientation and configuration of water dimer and trimer on Pt(111). A, B and C planes are defined such that the side views of the structures could be easily identified. The Pt, O and H atoms are represented by white, red and black spheres, respectively. Hydrogen bond donor molecules are signified by a white cross. Reprinted with permission from [20]. Copyright 2010 by Elsevier

to the adsorbed monomer. The bond angle of the donor molecule is wider than the bond angle of the acceptor molecule. Such observation is attributed to the possible charge transfer from the donor to the surface and the back donation of charge by the surface to the acceptor molecule.

The adsorption of a water trimer and other clusters with greater number of H_2O molecule (e.g. hexamer) on metal surfaces have also been studied theoretically [22]. Specifically, a cyclic arrangement of water clusters was found to exist on Pt(111). In such arrangement, one of the H atoms of H_2O forms bond with another molecule while the other H atom is free from interaction. It is further found that the molecules are almost planar with reference to the surface plane. Also, in the trimer case, all the water molecules in the cyclic configuration have almost the same O–Pt distances, O–H bond lengths, and HOH angles. The parameters are greater than the corresponding parameters in an adsorbed monomer. The adsorption energy per molecule is larger than the monomer case, but is smaller than the dimer case. Notably, a separate investigation revealed that the adsorption energy per water molecule increases with increasing cluster size [20]. The ring-like structure of a trimer is not the most stable adsorption configuration on the surface. An open or non-cyclic (as shown in Fig. 2.10) adsorption structure is more stable than the ring-like arrangement. Furthermore, the water molecules have comparable O–Pt distances in the cyclic structure. In the open structure, one of the molecules is closer than the other two molecules with characteristics similar to the dimer case. This orientation depicts the presence of donor and acceptor water molecules on the surfaces. Such adsorption properties rationalize the difference in the stabilities obtained for these different structures for the trimer case.

The adsorption of water molecules on stepped metal surfaces is of particular interest due to the known reactivity of the metal atoms in the stepped region. In such case, it is possible to achieve a chain-like structure of water molecules that follows the arrangement of the surface atoms at the stepped site. Such chain structures of water molecules are present in biological systems. Thus, elucidating this arrangement and its corresponding interaction with other systems is important not only in fundamental science, but also to its applications in various fields.

Figure 2.11 shows the structures of a one-dimensional water chain at the step region of a Pt(322) surface. The chain is formed by the hydrogen bonds between each molecule. With reference to the step surface, a molecule could orient into two configurations. First, the H atom of the molecule that is not interacting with another water molecule is pointing inward the step (marked as "1" in Fig. 2.11). Second, in the "2" structure, such H atom is pointing outward the step. These configurations could be referred to as H-in and H-out, respectively. It is essential to compare the strength of adsorption of a monomer on the steps region of the surface, with respect to these possible structures. In general, the molecule prefers to be adsorbed at the on-top site of the step region. The adsorption of a single water molecule on the step is more stable in the H-in structure than the H-out structure, with adsorption energies of -0.45 eV and -0.43 eV, respectively [22]. These adsorption energies are larger than in the case of a monomer on flat Pt(111) (-0.29 eV). Note however, that the difference in the H-in and H-out adsorption energies is barely significant.

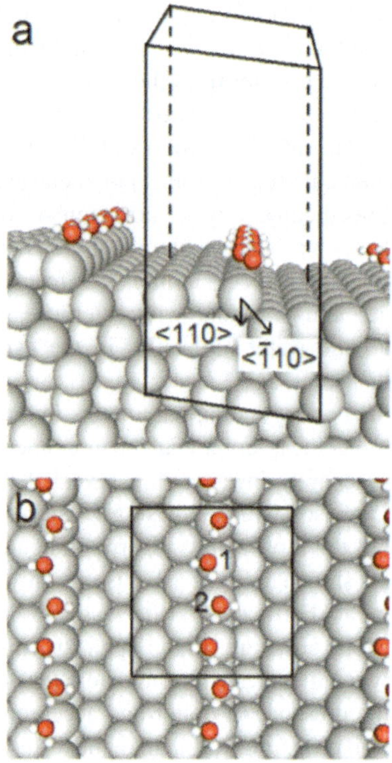

Fig. 2.11 **a** Side and **b** top views of the adsorption structure of a 1-dimensional water chain at the step region of the Pt(322) surface. Reprinted with permission from [22]. Copyright 2004 by the American Physical Society

Thus, it can be inferred that while the adsorption energy of water on steps is larger than on terrace, the orientation of the molecule could assume either the H-in or H-out structure. On the other hand, a study revealed the stronger adsorption of H_2O on kink sites than on the step region [20]. The calculated adsorption energy is about 0.20 eV larger than on the terrace region. In the study, two different kink sites of Pd(763) and Pd(854) surfaces were considered. These surfaces differ such that they have (111)- and (100)-oriented edges, respectively. The adsorption structures of the monomer on these sites are shown in Fig. 2.12. The monomer assumes the H-in structure on Pt(854) and H-out structure on Pt(763). The enhancement in the adsorption energies on step and kink sites is attributed to the modified electronic properties of the surface atoms in those regions.

Given such configurations, it is possible to investigate the chain structures of water with all molecules in H-in or H-out orientation, and a configuration where the H-in and H-out structures alternately form the chain (Fig. 2.11b). For the 1-D chain, the alternate/zigzag arrangement is favored as compared to chain structures that are mainly composed of H-in or H-out structures. The calculated adsorption energy per water molecule for the chain with alternate H-in and H-out structures is −0.48 eV. On the other hand, the adsorption energies for the chains with H-in and H-out structures are −0.43 eV and −0.39 eV, respectively. More importantly, it was generally observed that the adsorption energy per water molecule increases with

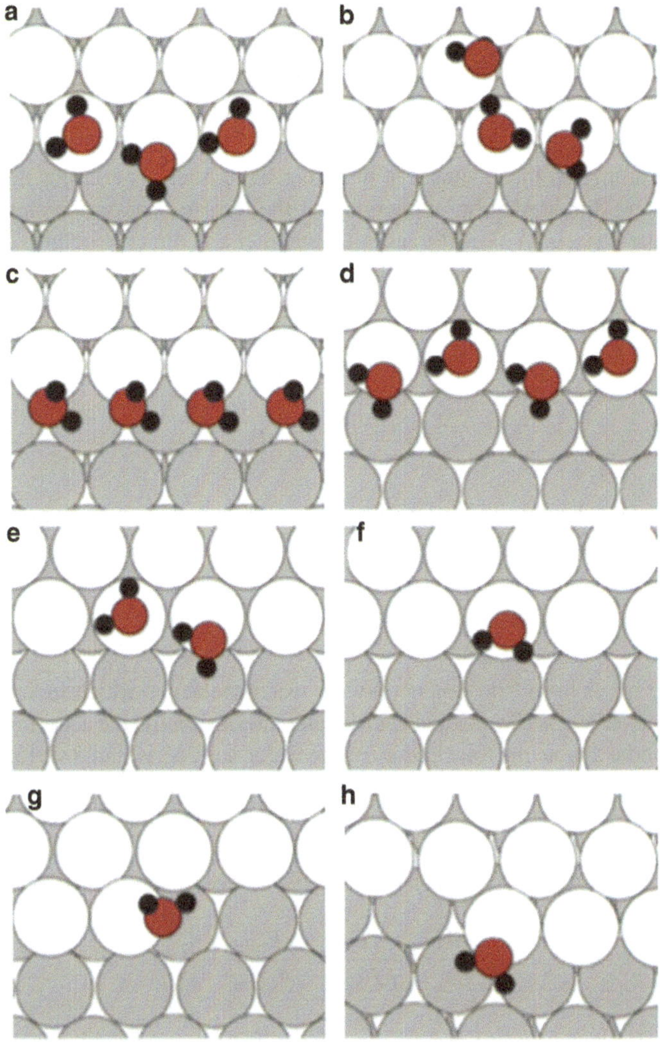

Fig. 2.12 Adsorption structures of water molecules on step and kink regions of Pt surfaces. Reprinted with permission from [20]. Copyright 2010 by Elsevier

increasing cluster size. A large increase in the adsorption energy exists between the monomer and dimer and slightly increases with additional water molecules. Note that the increase in the adsorption energy is more prevalent in adsorption on the terrace site than on step site.

In order to understand this trend in the adsorption energies with number of adsorbed water molecules and its dependence on the adsorption sites, it is necessary to determine the contributing factors to the calculated adsorption energies. In water cluster adsorption, the effect of hydrogen bond between the molecules on the calculated adsorption energy should be considered. It is therefore necessary to quantify the strength of the hydrogen bond on cluster adsorption. For the case of a monomer, the adsorption energy of water is solely attributed to the strength of the surface–water interaction. Thus, the stronger adsorption on the step and kink regions than on the terrace region is mainly due to the different electronic structures of surface atoms on those sites. On the other hand, there is a larger difference in monomer and dimer adsorption energies on the terrace than on the step sites. For example, there is a 0.15 eV difference on monomer and dimer structures on Pt(111), but only 0.07 eV difference is calculated for the same structures on the step site of a Pt surface. Such different results are clarified by quantifying the hydrogen bond strength (E_H) and by getting the ratio of adsorbate–adsorbate to substrate-adsorbate energies (AA/SA). These quantities are defined as follows: [20]

$$E_H = \frac{n}{n_H}\left(E_{ads} - E_{ads,m}\right) \tag{2.2}$$

where $n_H = n - 1$ for finite cluster and $n_H = n$ for extended/chain or cyclic clusters. Here E_{ads} is defined as

$$E_{ads} = \frac{1}{n}\left[E_{tot,ads} - \left(E_{surf} + nE_W\right)\right]. \tag{2.3}$$

E_{ads}, therefore, is the adsorption energy per water molecule, which incorporates the contributions of surface-water and water-water interactions on the adsorption. $E_{tot,ads}$, E_{surf}, E_W and n denote the total energy of the water-surface, isolated surface, isolated water systems and the number of water molecules on the surface, respectively. $E_{ads,m}$ defines the adsorption energy of a monomer on the surface under consideration. E_H, therefore, defines the average of hydrogen bond strength of water clusters and chain on a surface. The AA/SA factor, on the other hand, is calculated by using the equation

$$\frac{AA}{SA} = \frac{n_H E_H}{nE_{SA}} = \frac{E_{ads}}{E_{ads,m}} - 1 \tag{2.4}$$

where E_{SA} is given as $nE_{SA} = nE_{ads} - n_H E_H = nE_{ads,m}$. The dependence of E_{SA} on the adsorption energy of a monomer is based on the idea that it is only on the monomer case in which the actual influence of the surface toward the adsorbate could be obtained.

On flat Pt surface, the obtained values of E_H and AA/SA ratio for the dimer are −0.30 eV and 0.50 respectively. On the stepped surface, the values are −0.14 eV and 0.15, respectively. For the trimer case, E_H and AA/SA are −0.27 eV and 0.60 on the flat surface, respectively. The corresponding values of E_H and AA/SA on the stepped surface are −0.15 eV and 0.22, respectively. It can be seen that there are larger values of E_H and AA/SA on the flat surfaces for both the dimer and trimer as compared to the case of stepped surfaces. The large values on the flat surface indicate that hydrogen bond strength is responsible for the stabilization of dimer and trimer on the surface. On the other hand, the lower values on stepped surfaces indicate that the stabilization is mainly due to the reactivity of the surface atoms in the stepped region with a weaker hydrogen bond contribution as compared to those of the flat surfaces. This reactivity is likewise evident to the stronger adsorption energies of water monomer on steps and kinks than the adsorption on flat surfaces.

Despite the undeniable value of DFT-based calculations in understanding the interaction of water on metal surfaces, the widely used GGA functionals such as PBE or PW91 incorrectly predict that none of the experimentally characterized icelike wetting layers on metal surfaces are thermodynamically stable [23]. This challenges the role of DFT in water adsorption studies and casts doubts in the structural characterization of water overlayers on metal surfaces. Carrasco et al. explored the critical role of van der Waals dispersion forces on this system to shed light on such long-standing problem of metal wetting [24]. Comparing the PBE functional with a optB88-vdW nonlocal functional to study the water–ice adsorption systems on Cu(110) and Ru(0001), it was found that nonlocal correlations contribute substantially to the water-metal bond, and that this factor is crucial to the relative stabilities of wetting layers and 3D bulk ice. Figure 2.13 shows the adsorption energies of model water overlayers on Cu(110) and Ru(0001) compared to the lattice energy of bulk ice *Ih*. The first observation is that PBE predicts bulk ice to be more stable than the water overlayer structures on Cu(110) and Ru(0001) surfaces, which is the incorrect

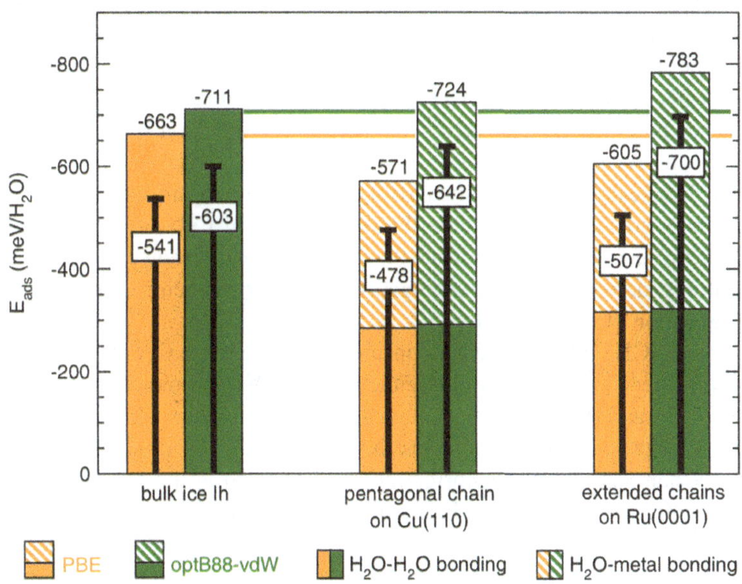

Fig. 2.13 The adsorption energies of water overlayers on Cu(110) and Ru(0001) compared to the lattice energy of bulk ice *Ih* using PBE and optB88-vdW functionals. The adsorption energy is decomposed into water-water and water-metal bonding. The thin centered black bars and associated numbers are the adsorption or lattice energies corrected for zero point energies. Reprinted with permission from [24]. Copyright 2011 by the American Physical Society

preference for bulk ice reported in DFT studies that utilized the PBE and PW91 functionals. Interestingly, the optB88-vdW functional gives the correct energy ordering, predicting that the overlayer is marginally more stable than bulk ice, in agreement with observations in experiments. The dispersion forces between water and metals are sufficiently large to favor the formation of 1D and 2D water structures on surfaces over the 3D bulk ice. This arises from the much larger polarizability of the metal atoms compared to oxygen and hydrogen. Their results highlighted the importance of dispersion correction to DFT calculations in the study of water on metals.

2.3 Methane Activation on Metal Surfaces

Methane (CH_4) is an important chemical that is used in industries specifically for synthesis gas (syngas) production. Syngas produced from methane is rich in CO and hydrogen, which is useful in the production of valuable compounds such as methanol, aldehyde and ammonia. This importance of methane is one the major reasons of the numerous studies related to its interaction with materials. The process of converting methane to syngas is known as methane reforming. In steam methane reforming (SMR), methane reacts with water to form CO and H_2. From such reaction, it is recognized that methane is a promising source of hydrogen gas, which could be utilized in hydrogen fuel cell applications. The remaining CO from syngas is usually allowed to react with water to form CO_2 and H_2 to further increase the yield of hydrogen gas. Other methods of methane reforming include the dry reforming reaction, where CH_4 reacts with CO_2, and partial oxidation with O_2 gas.

A key elementary step to realize SMR is the dissociation or dehydrogenation of CH_4. Such process requires large amount of energy, requiring the need for efficient and cheap catalysts for practical realization. In this regard, this section focuses on previous works that investigated the dissociation of CH_4 on different flat and stepped metal surfaces, which indicate the impact of the different environment of the surface atoms (depending on the corrugation of the surface being considered) on the dehydrogenation of the molecule.

In general, CH_4 adsorbs weakly on metal surfaces. On Cu(111) for instance, its adsorption energy is almost negligible (−0.01 eV). On more reactive surfaces such as Pd, Pt and Ni, its adsorption energies are −0.10 to −0.20 eV [25]. This range of adsorption energies indicates that the molecule is physisorbed on metal surfaces. When van der Waals corrections are included in the calculation, greater adsorption energies for CH_4 can be found [26, 27]. The magnitudes of adsorption energies of CH_4 decomposition fragments are higher than that of CH_4 molecule. For example, the adsorption energies of CH_3, CH_2, CH, and C are −1.465 eV, −3.908 eV, −5.149 eV, and −5.112 eV, respectively, which are calculated with reference to the unstable gas-phase CH_x ($x = 0$–3) fragments. These decomposition fragments are more stable on the surface than the case of CH_4 due to their capacity to form bonds with the metal surface.

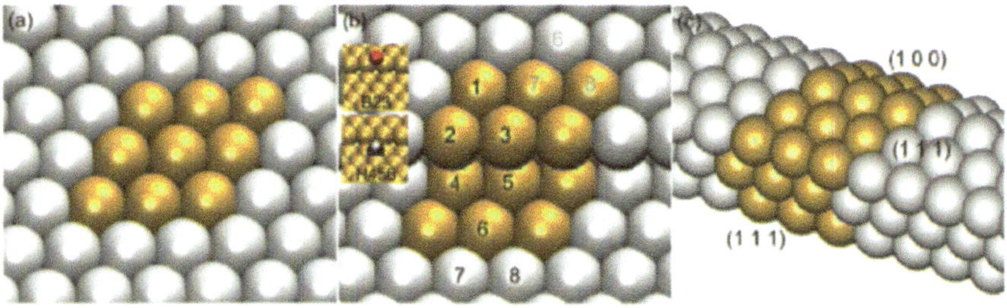

Fig. 2.14 Surface models of different flat and stepped Rh surfaces: **a** Rh(111), **b** stepped Rh(211) and **c** Rh nanorod with (111) and (100) facets edges. Reprinted with permission from [28]. Copyright 2011 by the American Chemical Society

Rh surface is one of the most commonly used catalysts for SMR applications. It was found in previous works that it is an active catalyst for CH_4 decomposition. DFT-based calculations were performed to investigate the dissociation paths of the molecule on different Rh surfaces. Notably, Grootel et al. compared the performance of different Rh surfaces (flat and stepped surfaces) for such process [28]. The dehydrogenation of CH_4 was studied on Rh(111), Rh nanorod with (111) facet and edge regions, and on stepped Rh(211) surfaces, which are shown in Fig. 2.14. Their study evaluates the reactivity of Rh atoms toward CH_4 dehydrogenation, in relation to their coordination numbers. In other words, it demonstrates how the behavior and dissociation of CH_4 and its fragments depend on the environment of the Rh atom on the surface.

The CH_x fragments are most stable on the stepped Rh(211) surface except for CH_2, which is found to be most stable on the edge of the Rh nanorod. CH_2 adsorbs on the bridge site of the edge region of the Rh nanorod with an adsorption energy of −5.29 eV, which is about 0.06 eV more stable than on the case of stepped Rh(211) surface. The trend further shows that the adsorption energies increase as the number of H atoms that is being detached from CH_4 increases. In such case, the adsorption energies for CH_3, CH_2, CH and C are −3.05 eV, −5.23 eV, −7.61 eV, and −9.20 eV, respectively. The corresponding adsorption energies on flat Rh(111) are −2.73, −5.14, −7.57, and −8.79 eV, respectively. Note however that the reference energies for these adsorption energies involve the CH_x (x = 0–3) fragments, which are unstable gas-phase species. Also, the adsorption energies of CH_x fragments are significantly influenced by the adsorbate coverage. Weaker adsorption is observed for higher adsorbate coverage, signifying the repulsive interaction between the adsorbates [28, 29]. The observed stronger adsorption energies on the corrugated surfaces are clearly due to the lower coordination of the Rh atoms on those surfaces as compared to the surface atoms on the flat surface regions. As explained in the earlier sections, the surface atoms with lower coordination are more reactive than the surface atoms with higher coordination.

Figure 2.15 shows the reaction path of CH_4 dissociation to C atom and four H atoms on different flat Rh surfaces—Rh(111) with 0.11 ML and 0.25 ML coverages of CH_4, and on the flat (111) facet of Rh nanorod. It shows that the reaction energies for the system with lower CH_4 coverage are more stable than the system with higher

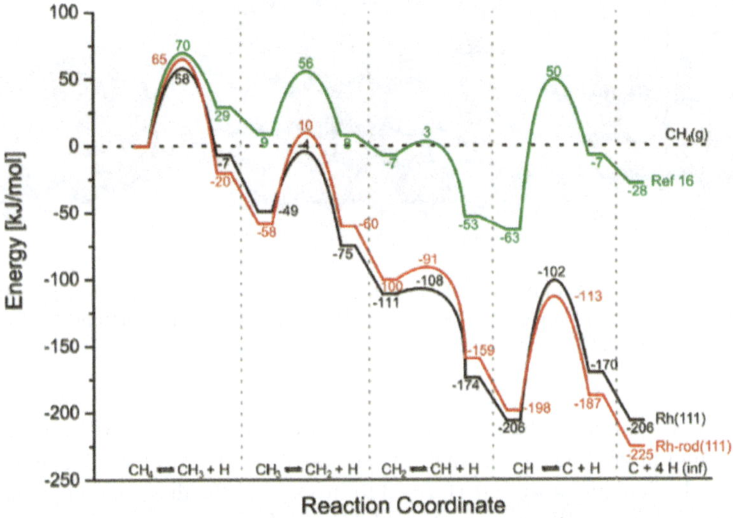

Fig. 2.15 Potential energy diagram of CH_4 dehydrogenation to C and 4 H on Rh(111) (0.11 ML of CH_4) (black), (111) facet of Rh nanorod (red), and on Rh(111) (0.25 ML of CH_4). Reprinted with permission from [28]. Copyright 2011 by the American Chemical Society

CH_4 coverage. Again, this signifies the presence of repulsive interaction between the adsorbates. Such interaction likewise results to larger activation energies for the dissociation reactions that involve higher CH_x adsorbate coverages than on cases with lower adsorbate coverage. Also, the adsorption structures of the molecules strongly influence the resulting activation barrier in the dissociation process. For instance, the activation barrier for CH_3 dissociation into CH_2 and H is larger on Rh nanorod than on Rh(111). On the nanorod, the H atoms of CH_3 are pointing outward from the surface atoms and the diffusing/detaching H atom should overcome a large distance as compared to the case of Rh(111). On the surfaces considered, CH dissociation encounters the largest barrier. Because CH adsorbs perpendicularly to the surface plane, bending the molecule before the abstraction of H requires large energy. Such phenomenon was also observed for the case of CH_4 dissociation on Cu(111) and on other metal systems [30, 31].

It is important to point out that most of thedissociation processes involve H atoms that move over the top site of the Rh surfaces. This signifies that the electronic structure of the Rh atom where the dissociation occurs affects the resulting activation energies. Such effect is dominantly observed on the dissociation of CH_4 on surfaces with steps and edges. The Rh atoms in the step and in the edge regions have lower coordination number as compared to the atoms in flat surfaces. Figure 2.16 shows the potential energy diagram of CH_4 dehydrogenation on Rh(211) stepped surface and on the edge region of the Rh nanorod. As compared to Rh(111), the reaction energy is more negative on Rh(211) than on flat Rh(111) as shown in Fig. 2.15. This signifies the more stable adsorption of the decomposed species of the molecule on the stepped surface. While the adsorption of CH_4 and its dissociated species is

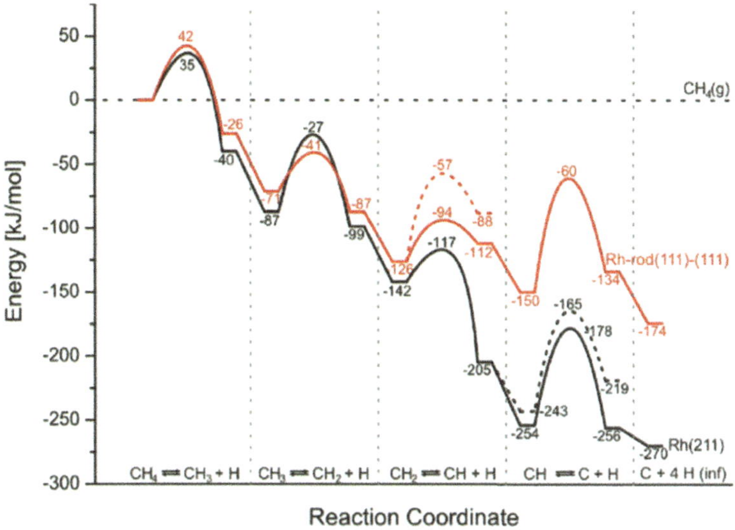

Fig. 2.16 Potential energy diagram of CH_4 dehydrogenation to C and 4H on stepped Rh(211) surface (black) and on the edge region of Rh nanorod (red). Reprinted with permission from [28]. Copyright 2011 by the American Chemical Society

stronger on the stepped surface, there is no obtained trend with regard to the calculated activation barriers. On Rh(211), CH_4 dissociation into CH_3 and H is accompanied by 0.36 eV activation barrier. The corresponding barrier on Rh(111) is 0.60 eV. Lower barrier for CH dissociation is also found on the stepped surface which has a value of 0.79 eV as compared to the 1.08 eV barrier on flat Rh(111). In such cases, the lower barriers could be attributed to the reactive nature of the Rh atoms in the steps resulting to the lowering of the activation energies. For CH_3 and CH_2 dissociation reactions, however, the reported activation energies are lower on the flat Rh(111) than on the stepped Rh(211) surface. Here, the adsorption structures of the molecules significantly influence the activation barrier. Due to the corrugated (211) surface, the CH_3 and CH_2 molecules are adsorbed differently as compared to the adsorption structures on the flat (111) surface. The CH_2 fragment on Rh(211), for instance, needs to rotate from its initial adsorption configuration to activate the dissociation process. On the other hand, the CH_2 fragment on Rh(111) does not require such process, leading to a lower activation barrier. This is also the case for the lower activation barrier for the dissociation of CH_3 on the edge of the nanorod (0.31 eV) as compared to the dissociation on the stepped surface (0.62 eV). In such cases, CH_3 is already in the activated state on the former while it needs to rotate first before dissociating on the latter. Nonetheless, the large difference in reaction energies can be seen, in which the C and four H atoms are more stable on the stepped surface than on the edge of the nanorod.

The dissociation of CH_4 on different Rh surfaces demonstrates the more reactive nature of atoms in the steps than in the flat surfaces. However, there are cases in which the geometry of the molecule is a factor that influences the barrier, as in the case of CH_3 dissociation. The dissociation of CH has the highest activation barrier for all the considered surfaces. It suggests that regardless of the surface corrugation of the Rh surfaces, the energetics for CH_4 dehydrogenation could be similar. The calculated *d*-band centers for the different Rh atoms further support the calculated reactivity toward CH_4 dehydrogenation and adsorption strength of the CH_x fragments on the different surfaces.

These discussions tackled how the coordination number of surface atoms influences the dissociation of CH_4. In materials design, this suggests that the surface with atoms that have very low coordination number is desired to promote the easy dissociation of CH_4. In such case, an adatom on a surface would be the most practical surface design. The group of Kokalj investigated the dehydrogenation of methane on different Rh surfaces: flat Rh(111), Rh surface with steps, Rh surface with added row (ad-row), and a surface with Rh adatom [32]. The coordination number for these systems are 9, 7, 5 and 3, respectively. The surfaces are shown in Fig. 2.17, while the corresponding adsorption energies and activation barriers for CH_4 and CH_3 dehydrogenation are shown in Fig. 2.18a, b. Results shown in Fig. 2.18a demonstrates the stronger adsorption of the species with decreasing coordination number, although the H adsorption energies are almost similar for all the surfaces considered. Also, it shows the weaker adsorption of CH_3 and CH_2 on the Rh adatom than on the ad-row surface despite its lower coordination number.

The dissociation of CH_4 into CH_3 and H obeys the expected trend. The activation energy decreases as the coordination number decreases, signifying the good reactivity of the more isolated surface atoms. However, the activation energy for CH_3 dissociation into CH_2 and H is highest for the adatom case. Analysis of the geometries of the dissociating CH_3 in such case shows that the CH_3 and CH_2 fragments prefer to bond with at least two Rh atoms. Such preference is the origin of the lower adsorption energy of the species on the adatom system. The complex dissociation of CH_3 into CH_2 and H, which involves the diffusion of H across the adatom before it reaches the adatom-to-surface bridge site, explains the high activation barrier. Due to this high activation barrier, subsequent reaction or the full dehydrogenation of the molecule may be suppressed. These findings suggest that while the lower coordinated surface atoms could indeed result to lower activation energies, the complex interaction of

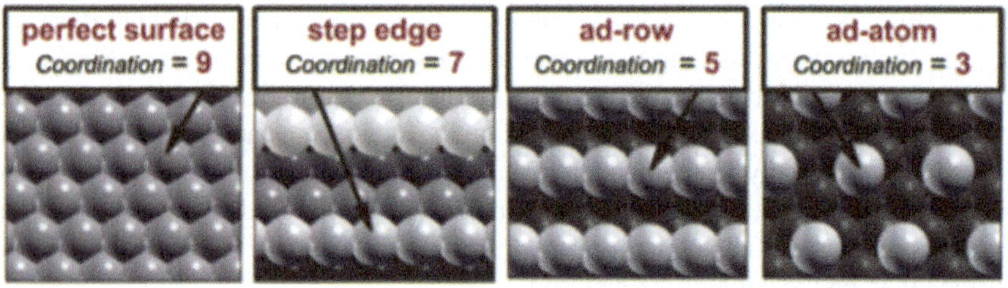

Fig. 2.17 Geometric structures of Rh(111) (from left to right): perfect surface, (100) step-edge, ad-row, and ad-atom. Reprinted with permission from [32]. Copyright 2004 by the American Chemical Society

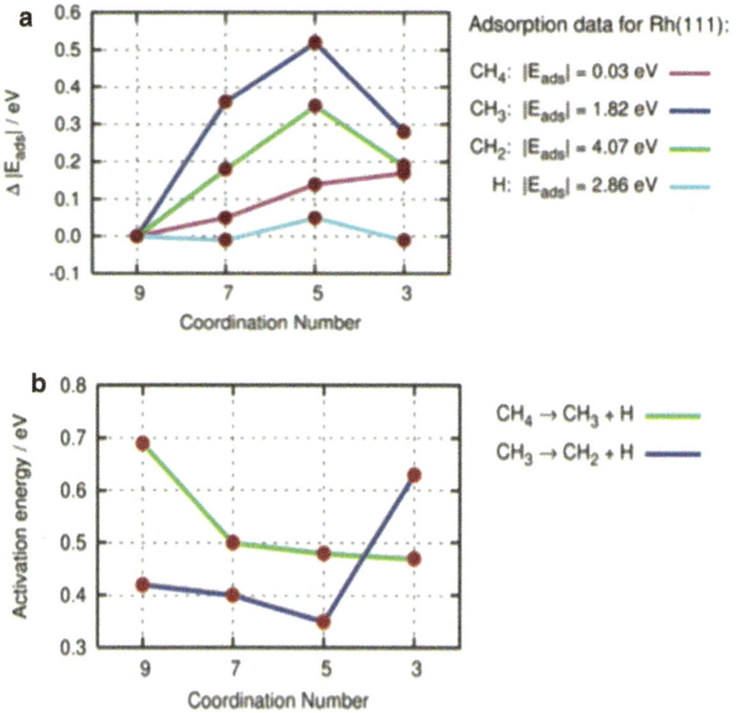

Fig. 2.18 **a** Adsorption energies of CH_4, CH_3, CH_2, and H on the different Rh surfaces with different coordination numbers of surface atoms. **b** The activation energies for CH_4 and CH_3 dissociations on the different Rh surfaces with different coordination numbers of surface atoms. Reprinted with permission from [32]. Copyright 2004 by the American Chemical Society

the molecules with the surface has strong influence on the desired molecule-surface process. It is necessary to carefully examine the interactions between adsorbates and surfaces despite the occurrence of trends in these surface related processes.

While DFT studies on the interaction of methane on metal surfaces provide significant insights into the activation of its C-H bonds, a more realistic model takes into account the oxidized state of the metal catalyst. This is particularly important for the combustion of methane on Pd-based catalysts. Under oxygen rich conditions and temperatures below 950 K, Pd exists in the form of an oxide [33, 34]. In situ XRD and XPS measurements showed that PdO(101) develops preferentially during the oxidation of Pd(100) [35–37]. This oxide formation coincides with the increased rates of methane oxidation, suggesting PdO formation to be responsible for the exceptional activity of Pd-catalysts for the catalytic combustion of methane [37–39].

Figure 2.19 shows the side and top views of the PdO(101) surface. Pd atoms on the surface are marked as either 3f or 4f to denote the threefold and fourfold coordinated Pd atoms, which are respectively referred to as Pd_{3f} and Pd_{4f} atoms. Pd forms a square planar coordination with O atoms as shown by the red and blue planes in Fig. 2.19. Pd_{3f} is coordinatively unsaturated (cus) with a coordination vacancy on top of the

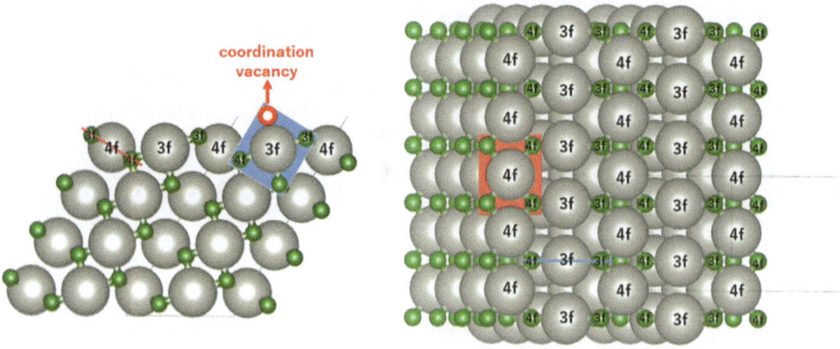

Fig. 2.19 The side (left) and top (right) views of the slab model for PdO(101) surface. Gray and green atoms represent Pd and O, respectively. Pd and O atoms on the surface are marked as 3f or 4f, indicating a threefold or fourfold coordination. The coordination plane for Pd_{3f} and Pd_{4f} atoms are shown by the blue and red planes, respectively [40]

Pd_{3f} atom. The coordination plane of the unsaturated Pd_{3f} atom is perpendicular to the surface, while that of the Pd_{4f} atom is tilted with respect to the (101) surface. Analysis of the electronic properties of PdO(101) points to the importance of Pd_{3f} in the activation of the methane C–H bonds. As shown in Fig. 2.20a, methane adsorbs at the coordinatively unsaturated Pd_{3f} atom on the surface, with two "activated" C–H bonds coordinated to Pd atom and the other two C–H bonds pointing away from the surface. The unoccupied d_{zz} state of the Pd_{3f} atom facilitates the electron donation from the filled σ molecular orbital of CH_4, while its occupied d_{yz} orbital provides a back donation to the unfilled σ anti-bonding CH_4 molecular orbital, resulting in the elongation and weakening of the Pd-coordinated C–H bonds. This mechanism is shown in Fig. 2.20b.

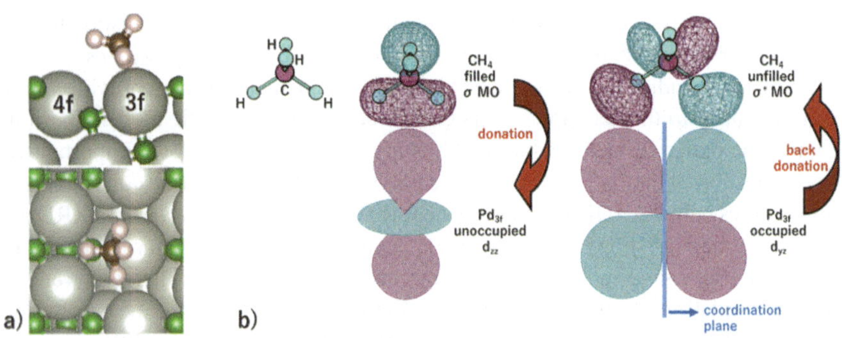

Fig. 2.20 **a** Optimal adsorption configuration of CH_4 on PdO(101) surface. **b** Mechanism of electron donation and back-donation between CH_4 and the coordinatively unsaturated Pd atom. Lobes of the same color denote the same phase of the wavefunction. Reprinted with permission from [40]. Copyright 2019 by The Royal Society of Chemistry

2.4 Adsorbate Induced Surface Segregation in Bimatellic Systems

Bimetallic systems are among the popular catalysts employed in industries due to their known advantages as compared to their monometallic components. Among these advantages includes the enhancement of the reactivity and selectivity of the system and the improvement of its mechanical stability. Such performance of the bimetallic materials is explained by the interactions of the monometallic components, which results to the modification of the electronic and geometric properties of the system. Thus, a careful investigation of such systems could lead to successful materials design that can provide solutions on issues related to efficiency, cost, and durability of devices for specific applications.

In bimetallic surfaces, one of the most important phenomena that should be taken into consideration is surface segregation. It is the surface enrichment of one type of atom due to the difference in the surface energies of the monometallic components. In PdAg surfaces, for example, it is more favored to have a higher concentration of Ag atoms at the topmost layer due to the lower surface energy of Ag than Pd. For systems composed of monometallic components with comparable surface energies, it is found that moderate segregation will take place. Such considerations are indeed important in computational simulations to address a more realistic model.

Ruban et al. investigated the trend in the segregation of clean bimetallic surfaces by performing density functional theory based calculations in conjunction with the coherent potential approximation (CPA) and the Green's function technique [41]. They found that the surface segregation process is strongly influenced by the difference in the surface energies of the monometallic components and of the crystal structure of the isolated component atoms. The surface segregation energy, which is the energy cost of moving an impurity atom to the topmost layer of the host surface, can be calculated as follows:

$$E^B_{segr-\lambda} = \left. \frac{dE_{surf}\left(A_{1-c_\lambda}B_{c_\lambda}\right)}{dc_\lambda} \right|_{c_\lambda=0}. \quad (2.5)$$

The surface energy of the system that is composed of an $A_{1-c_\lambda}B_{c_\lambda}$ alloy in the λth layer of a surface is defined as $E_{surf}\left(A_{1-c_\lambda}B_{c_\lambda}\right)$. B and A denote the impurity atom and the atoms of the host surface, respectively. The surface energy of the system is computed as

$$E_{surf}\left(A_{1-c_\lambda}B_{c_\lambda}\right) = \sum_{\lambda'}\left(E_{\lambda'} - E_{bulk}\right) - \mu c_\lambda. \quad (2.6)$$

$E_{\lambda'}$ is the total energy per atom of the λ'^{th} layer. The total energy per atom of the host and the chemical potential of the B atom in the host A are defined as E_{bulk} and μ, respectively. The chemical potential μ can be calculated as

$$\mu = \left. \frac{dE_{bulk}(A_{1-c}B_c)}{dc} \right|_{c=0}. \tag{2.7}$$

From these equations, a negative segregation energy indicates that the impurity prefers to segregate to the surface region of the host. On the other hand, a positive value signifies the preference of the impurity atom to stay in the interior of the surface.

The surface segregation energies of different transition metal bimetallic systems are shown in Fig. 2.21 as a 24 × 24 matrix. The red and blue colors signify negative and positive surface segregation energies, respectively. The matrix implies that with a small amount of Ag in Pd surface, Ag will most likely segregate to the topmost

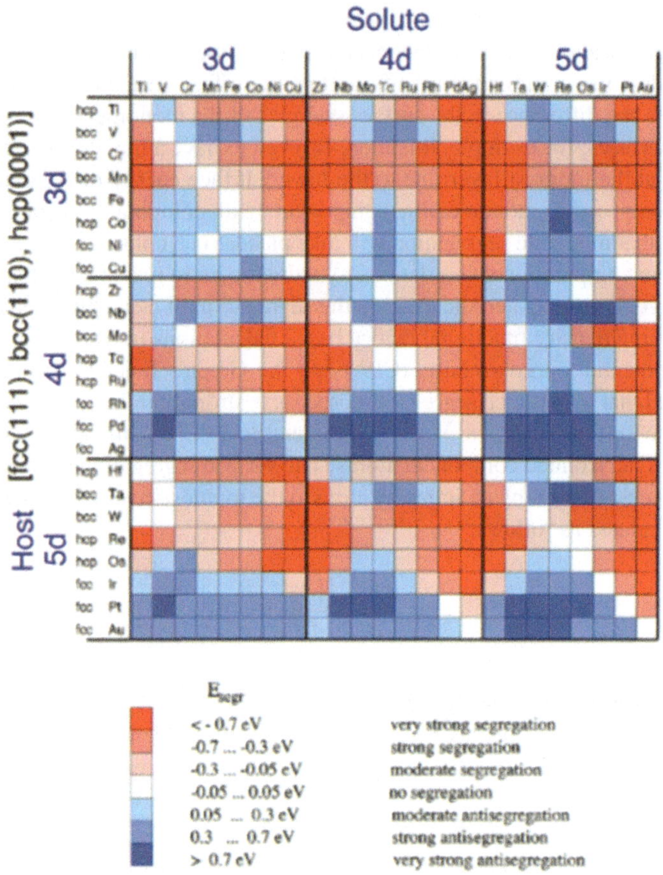

Fig. 2.21 Surface segregation energy matrix for different transition metal bimetallic systems. Segregation (red) implies that the solute/impurity atom will prefer to segregate to the topmost layer of the host surface. Antisegregation (blue), on the other hand, means that the solute/impurity atom will favor to stay in the interior of the host metal surface. Reprinted with permission from [41]. Copyright 1999 by the American Physical Society

layer. On the other hand, no segregation will be observed for Cu in host Pd surface. A solute Co in host Pd surface, classified as 'antisegregation', implies that Co will prefer to stay in the interior of the surface. The data provided is in good agreement with observations from experiments. Note that there are other models that also predict similar trends with regard to the segregation phenomenon in transition metals. One advantage of this presented model is that it captures the significant influence of the crystal structure of the component atoms, aside from the contribution of surface energy, to the segregation process. It was found that the dependence of segregation on the crystal structure is due to the valence *d*-electrons that are responsible to the local interatomic bonding in transition metals. In bimetallic system, the interaction between the host atom and the impurity atom somehow induces defect in the system. Such irregularity will reflect to both the electronic and geometric properties of the system. In DFT based calculations, these changes are included in the self-consistent field calculations.

Bimetallic surfaces with Pd as the host surface have garnered much attention in the past years due to the importance of Pd in catalysis, storage purposes, and as sensor devices. Specifically, for hydrogen storage and permeable membranes applications, Pd is alloyed with other metals to improve its mechanical stability and to increase its efficiency in allowing the diffusion of H atoms. Pure Pd surfaces allow the feasible dissociation of hydrogen molecules. However, the strong adsorption of the dissociated hydrogen atoms leads to their trapping on the surface region. Their diffusion or penetration is therefore hindered and is a usual problem in storage applications. In addition, hydrogen atoms in the bulk of Pd could cause the expansion of the material, which may result to the embrittlement of the host Pd. Alloying the material with other transition metals is among the best solutions for such issues. Through alloying, there is a chance to increase its bulk lattice constant and to allow easier diffusion or penetration of H atoms.

Løvvik investigated surface segregation in Pd(111) alloyed with different transition metals by performing DFT calculations [42]. In their study, a surface model composed of five atomic layers and four atoms in each layer is considered, as shown in Fig. 2.22. One transition metal impurity atom is introduced to predict its preference to segregate from the second subsurface layer (marked as 3) to the first subsurface layer (marked as 2) and to the topmost layer (marked as 1). Segregation energy is then determined by taking the difference in the total energies between the doped surface in which the impurity atom is in the topmost layer and in the first or second subsurface layer.

Figure 2.23 shows the calculated segregation energies of the different transition metals considered as impurity atoms in host Pd(111) surface. Here, the variation in the preference of the impurities to segregate or to remain in the bulk can be seen. Positive and negative segregation energies denote that the impurity will prefer to remain in the bulk and to segregate to the topmost layer, respectively. Consistent with the findings of Ruban et al., the impurity with lower surface energy than the Pd will segregate to the topmost layer [41]. Also, it was found that there is a correlation between the covalent radii of the impurities to the segregation energy. A large covalent radius will result to a more negative segregation energy. Thus, atoms with large radii will prefer

Fig. 2.22 Five-layer Pd(111) surface with four atoms per layer in the surface unit cell. The numbers 1, 2 and 3 denote surface atom in the topmost layer, first subsurface layer and second subsurface layer, respectively. Preference of the impurity atom (Ag, Au, Cd, Cu, Fe, Mn, Ni, Pb, Pt, Rh, Ru, and Sn) to segregate is investigated by substituting a Pd atom with the impurity atom. Reprinted with permission from [42]. Copyright 2005 by Elsevier

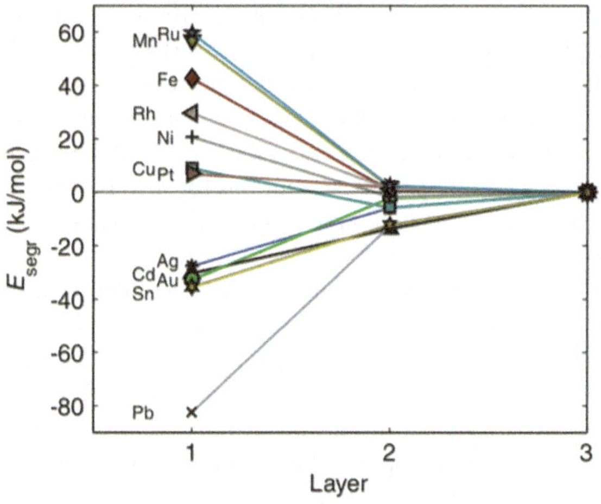

Fig. 2.23 Segregation energies of the different transition metals considered as impurity atoms in host Pd(111). The numbers 1, 2 and 3 denote surface atom in the topmost layer, first subsurface layer and second subsurface layer, respectively. Positive and negative segregation energies denote that the impurity will prefer to remain in the bulk and to segregate to the topmost layer, respectively. Reprinted with permission from [42]. Copyright 2005 by Elsevier

to segregate to the topmost layer. Consequently, the displacement or protrusion of the impurity atom with large atomic radius is outward (pointing away from the surface plane) with reference to the direction perpendicular to the surface. It is observed that the segregation is favored for impurity atoms with large outward displacement. It is concluded that the local distortions around the substituted atom could also be a factor in the segregation phenomenon.

The different behaviors of Ni and Cu as compared to other transition metals can also be seen in Fig. 2.23. Both atoms favor the second to first subsurface layer segregation as indicated by the negative segregation energies. However, the first subsurface layer to topmost layer segregation is not favored for both cases. These results suggest that Ni and Cu impurities will prefer to stay in the subsurface layer rather than to segregate to the topmost region. Such finding is consistent with the oscillatory concentration depth profile in PdNi systems and the moderate segregation in PdCu surfaces [43, 44].

The discussions so far have demonstrated that the process of surface segregation in bimetallic systems for the case of clean surfaces (i.e., surfaces with no adsorbate) strongly depends on surface energies, on geometric features of the component atoms, and on the relaxation of the atoms in the bimetallic surface. Interestingly, the process of reverse segregation can happen if adsorbates are present on the surface. This adsorbate-induced reverse segregation is attributed to the difference in the reactivities of the monometallic components toward different adsorbates. In PdAg for example, the Pd atoms will segregate to the topmost layer when an adsorbate (e.g. H atom) is introduced because the adsorption of H is stronger on Pd than on Ag. That is, the strong interaction of H with Pd drives the reverse segregation of Pd in PdAg systems.

Predictions on the tendency of adsorbate-induced segregation have been reported in the literature. Greeley and Mavrikakis studied the phenomenon by allowing the adsorption of H atom on different bimetallic transition metal surfaces [45]. The surface is constructed such that there is one layer of impurity atoms in the slab model. The layer resides either in the topmost layer or in the first subsurface layer. The preference of the layer to stay in the topmost or in subsurface layer is determined by calculating the segregation energy (E_{seg}) as demonstrated in an earlier work [41]. Aside from solving for the segregation energy, the binding energies (BE) of H on the pure host metal and pure solute metal are also obtained. The difference in these binding energies is correlated with the segregation energy to estimate the tendency for the adsorbate-induced segregation phenomenon. They constructed a model based on the determined segregation energies and H binding energies to predict which of the bimetallic surfaces will have the tendency to segregate in the presence of H.

Greeley and Mavrikakis also found that the adsorption energy of H is generally weaker on the considered surface alloys as compared to pure metal surfaces. Furthermore, there was a lower dissociation barrier of H_2 on the surface alloys than on the pure metal systems. This somehow deviates from the usual trend in reactions in which stronger adsorption of product species corresponds to lower activation barrier. In such case, one would presume that the weaker binding energy of H in surface alloys would mean higher dissociation barrier for H_2. An analysis of the obtained H_2-surface alloys transition states (TS) suggests that there is an extra stabilization of the TS geometry due to the interaction of the σ_{u*} orbital with the d_{xz} and d_{yz} metal states. Stabilization is also observed due to the interaction of the σ_g with d_{zz} orbital. Such stabilizations are results of (1) geometric changes in the surface alloys systems which provides deviations on the transition states and of (2) the possible decreased Pauli repulsion of the adsorbates with the s and p metal states. The first aspect is evident on the occurrence of molecular hydrogen precursor on the reaction path, which is not common on the dissociation path of H_2 on noble metals. The second aspect, on the other hand, is neglected in a previous model [46], which could be of relevant contribution in surface alloyed systems.

As earlier mentioned, the adsorbate-induced segregation in bimetallic surfaces is strongly attributed to the strength of the interaction of the adsorbate with the monometallic components of the system. In this regard, previous works sought to identify possible descriptors based on the electronic properties of the surfaces with the tendency to undergo the segregation process. Since the reactivity of pure metals is strongly dictated by the d-band center theory, similar attempt was made for the process of segregation in bimetallic surfaces in the presence of adsorbates [47]. Menning and Chen constructed a model that correlates the segregation energy of bimetallic systems with the d-band center. They showed that the segregation energy has better correlation with the d-band center if one will consider the occupied states of the d-band instead of treating the entire d-band (Fig. 2.24). The thermodynamic potential for segregation, ΔE^A_{seg}, of the impurity metal in the presence of adsorbate A is calculated using the equation

$$\Delta E^A_{seg} = \frac{E_{A/M-H-H(111)} - E_{A/H-M-H(111)}}{M}. \quad (2.8)$$

In this equation, $E_{A/M-H-H(111)}$ and $E_{A/H-M-H(111)}$ are the total energies of the adsorbate-metal surface system, in which, the bimetallic surface has overlayer impurity and subsurface impurity configurations, respectively. The nomenclature refers to M as an admetal layer and H as a host metal layer in the (111) surface facet. The M in the denominator indicates the number of host-admetal pairs per surface cell.

The figure suggests that for such systems, the correlation is better if the occupied states of the d-band are used for the d-band center calculations. Note that for the adsorbate induced segregation process, segregation takes place when the adsorbate is already present on the bimetallic surface. Thus, processes such as donation and back donation of charge involving the adsorbate and surface atoms could be neglected. With this assumption, the unoccupied states of the d-band could already be neglected since these are not significantly involved in the adsorbate induced segregation. Notably, a linear correlation was obtained between the d-band center and the segregation energy using this assumption. The results also suggest that for the bimetallic surfaces considered, segregation is more possible with O as an adsorbate as compared to the case of H. By investigating the adsorption of different adsorbates (H, Se, S, C, O and N), it was observed that the adsorbate induced segregation phenomenon strongly depends on the Pauling electronegativity of the adsorbate.

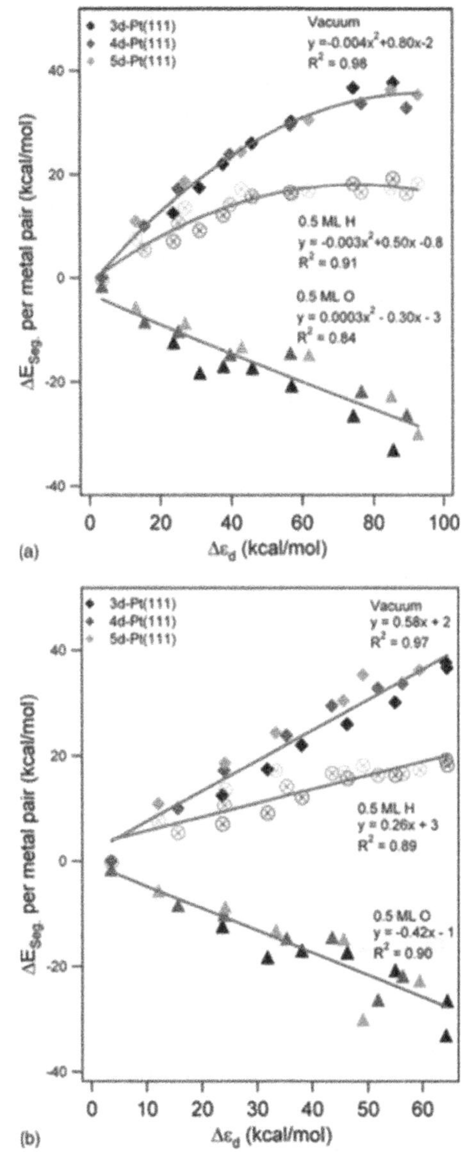

Fig. 2.24 Relationship between the segregation energy with d-band center of different bimetallic surfaces that considers the **a** full *d*-band center and the **b** occupied states of the *d*-band. Reprinted with permission from [47]. Copyright 2009 by the American Institute of Physics

Adsorbates with low electronegativities adsorb weakly on the surfaces, which results to a lesser chance of inducing segregation. On the other hand, adsorbates with high electronegativities adsorb more strongly on surfaces which could result in the possibility of inducing segregation between a subsurface admetal atom and a surface host atom.

The effect of surface segregation in bimetallic systems also has an impact on thediffusion of H atoms from the surface to the subsurface region. In a recent study, the absorption of H atom from surface to subsurface regions of pure Pd(110) and $Pd_3Ag(110)$ surfaces was investigated by employing DFT-based calculations [48]. The Pd_3Ag surface alloy is a stable alloy of Pd and Ag, which has been experimentally employed as hydrogen separation and hydrogen permeation membranes. It was observed that the permeation is better in the alloy system than in the pure Pd [49–51]. Thus, aside from the improved mechanical stability of the Pd_3Ag surface alloy, its performance toward hydrogen permeation is of profound importance in relation to its potential applications in hydrogen related reactions [52, 53].

The Pd_3Ag surface was constructed by cleaving the bulk alloy system in the (110) facet. Two possible ordered surface structures were obtained, in which, one surface is completely covered with Pd atoms while the other has two Ag and two Pd atoms in the topmost layer. Figure 2.25 shows these surface configurations composed of six atomic layers with four surface atoms in each layer. In this study, the adsorption structures of H atom on the surface and in the 1st and 2nd subsurface regions were first investigated. Then, the minimum energy path of the absorbing H atom was determined. In the subsurface regions, the H atom was found to reside in the tetrahedral and octahedral interstitial sites, which are its stable sites in the bulk and pure Pd surfaces as determined in previous works [54–58].

Figure 2.26 shows the reaction paths of an H atom absorption in Pd(110), (020202)-$Pd_3Ag(111)$, and (202020)-$Pd_3Ag(111)$ surfaces. The integers in the parentheses indicate the number of Ag atoms in each layer with the first integer referring to

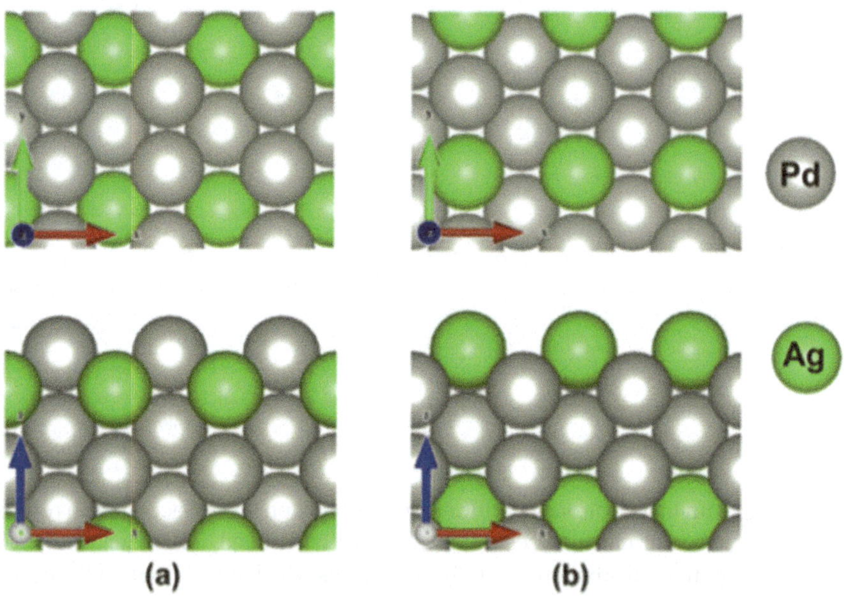

Fig. 2.25 Surface structures of **a** (020202)-$Pd_3Ag(110)$ and **b** (202020)-$Pd_3Ag(110)$ surfaces. The integers in the parenthesis indicate the number of Ag atoms in each layer with the first integer referring to the first atomic layer. Reprinted with permission from [48]. Copyright 2013 by Elsevier

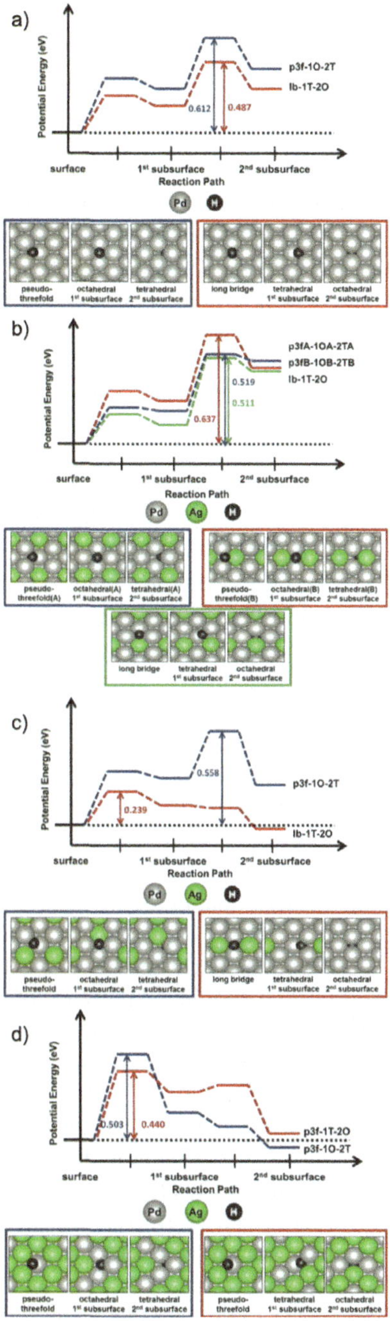

Fig. 2.26 Reaction path of surface-to-second subsurface layer H absorption in **a** Pd(110), **b** (020202)-$Pd_3Ag(111)$, **c** (202020)-$Pd_3Ag(111)$, and **d** (420000))-$Pd_3Ag(111)$ surfaces. The integers in the parenthesis indicate the number of Ag atoms in each layer with the first integer referring to the first atomic layer. The inset figures correspond to the geometric structures of H-surface systems in the surface and subsurface regions. Reprinted with permission from [48]. Copyright 2013 by Elsevier

the first atomic layer. In pure Pd surface, the potential energy shows an uphill energy profile as H moves from the surface to second subsurface region. It can be inferred from such profile and with the high activation barriers accompanying the absorption, the possibility of an H atom to be trapped in the surface region. Surface trapping of an H atom in Pd surfaces has also been observed in other studies [59]. The calculated activation barriers for H absorption in pure Pd(110) are 0.49 and 0.61 eV for the two possible paths considered in the study. In the (020202)-Pd_3Ag(110) case, the calculated activation barriers for H absorption are comparable with Pd(110). However, for the case of (202020)-Pd_3Ag, a low activation barrier is obtained if the H atom moves from the long bridge (lb) site to the subsurface region. The lb site in the surface is a bridge site formed by two Ag atoms. The lower activation barrier observed for this path can be attributed to the large interspacing of the surface atoms in Pd_3Ag and to the modified electronic structure of Pd atoms in the alloyed system. The bulk Pd_3Ag has larger lattice constant than Pd due to the larger atomic radius of Ag than that of Pd. With regard to electronic structure, Ag atoms transfer charge to the neighboring Pd atoms. This process further increases the reactivity of the latter toward H, which prefers to gain charge from the metal atoms during metal hydride formation.

The Ag atoms will prefer to segregate in the PdAg surface alloy due to its lower surface energy as compared to pure Pd. In this regard, the absorption of H was also investigated in the Pd_3Ag surface that is mainly composed of Ag atoms (420000). Figure 2.24d demonstrates that the barrier for this case is higher than in the Pd_3Ag(202020) case. The energy profile, nonetheless, is downhill and the effective activation barrier is only observed from the surface to first subsurface layer absorption. Similar to the downhill trend in the energy profile for H in (202020)-Pd_3Ag via the lb site, it can be inferred that the preference of H to go the subsurface is due to the more favored interaction of H with the subsurface Pd atoms. Given the variations in the calculated activation barriers for H absorption in different PdAg surfaces, it can be expected that the adsorbate-induced segregation process would play a crucial role in the amount of energy needed for reactions happening in the system.

It is demonstrated from the abovementioned studies that the surface atoms have a dynamic behavior in the presence of adsorbates. Aside from H atoms, this dynamic behavior of surfaces is more evident with larger molecules such as the hydrogenated compounds. In a previous work, it was shown that the effect of hydrogenated CO (COH and HCO) induced-segregation in Pd doped Cu surfaces is not only due to the stronger interaction between the molecule and Pd atoms, but is also influenced by the adsorption structure of the molecule [60]. For instance, HCO will favor the aggregated Pd atoms in Pd-Cu(111) regardless of the composition of Pd in the topmost layer and in subsurface layer. HCO adsorption on the surface is dependent on the available symmetry sites provided by the Pd atoms in the topmost layer. On the other hand, COH will favor more Pd atoms in the topmost layer for a more stable adsorption. This is due to the fact that COH will always prefer to adsorb on the hollow site, irrespective of the number of Pd atoms in the topmost layer. Thus, COH will choose to have more Pd atoms in the topmost layer to accommodate its adsorption on the hollow site.

Previous studies also elucidated the CO-induced adsorbate segregation in Pd-based bimetallic systems. In a clean Pd-Au(100) surface, Pd atoms will prefer to stay in the subsurface region. However, in the presence of CO, Pd will segregate to the topmost layer. The enrichment of Pd atoms in the surface region will then facilitate the dissociation of water molecules. Together with CO, and with the active dissociation of water on the surface, it was found that the Pd-Au bimetallic system is efficient for water–gas shift reaction [61]. Evidence of the adsorbate-induced segregation and the restructuring of bimetallic Pd-based surfaces in the presence of larger molecules (e.g. C_2H_2) were observed in other studies [62, 63]. These studies illustrate the complex behavior of bimetallic atoms in the presence of molecules, which implies that further investigations on the gas–bimetallic surface systems should be conducted after the synthesis or characterization of the clean surfaces.

2.5 Adsorbate Induced Reconstruction of Metal Surfaces

Another surface process that is equally important in metallic systems is the adsorbate-induced reconstruction. Surface reconstruction is observed when surface atoms move to other sites resulting in a completely different surface structure. An example of this is the reconstruction in fcc (110) surfaces, the least stable and the most open among the low-index surface facets of the fcc metals. The initially flat fcc(110) surfaces of Pt, Ir, and Au for instance, readily reconstruct into a missing-row reconstructed surface. The scanning tunneling microscope (STM) image of (1 × 2) missing-row reconstructed Au(110) surface is shown in Fig. 2.27a [64]. The structural image along the line I is shown in Fig. 2.27b. In the missing-row reconstruction, (111) microfacets are formed. The structures of unreconstructed fcc(110) and (1 × 2) missing-row reconstructed fcc(110) surfaces are shown in Fig. 2.28a, b, respectively. The inset in the figure

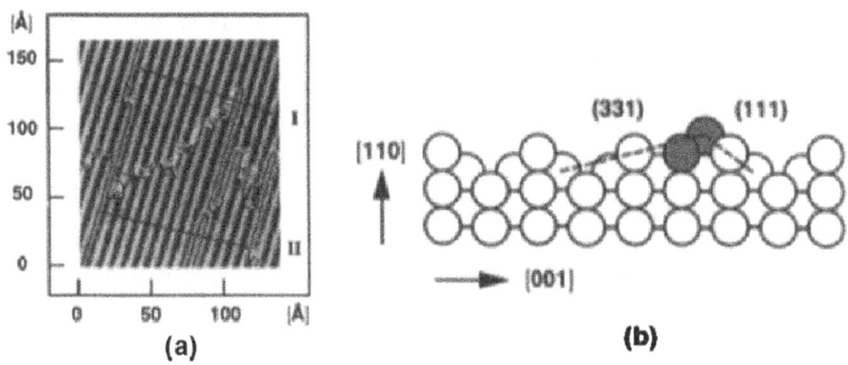

Fig. 2.27 **a** Scanning tunneling microscope (STM) image of (1 × 2) missing-row reconstructed Au(110) surface. The structural image along the line I of the reconstructed surface. Reprinted with permission from [64]. Copyright 1991 by Elsevier

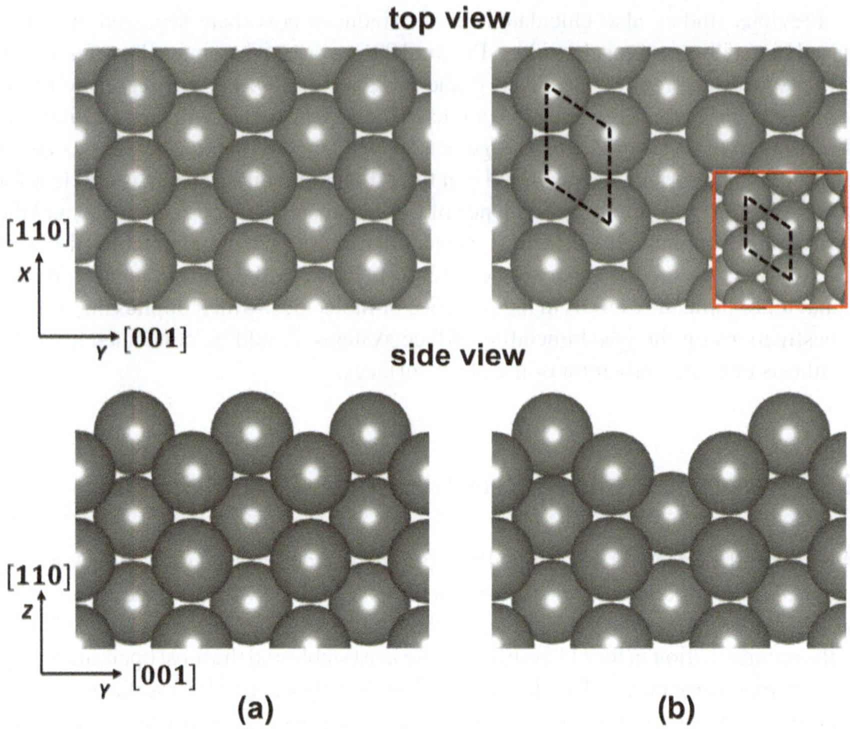

Fig. 2.28 Top and side views of the structures of the **a** unreconstructed fcc(110) and **b** (1 × 2) missing-row reconstructed fcc(110) surfaces. The inset in the **b** illustrates the formation of (111) microfacets in the reconstructed surface

illustrates the formation of (111) microfacets in the reconstructed surface. The (111) surface is the most stable and most compact surface facet of the fcc structure. It implies that the rearrangement of the surface atoms from an open structure into a more compact arrangement drives the reconstruction. The assumed missing-row structure is energetically more favored than the flat (110), which is the main reason for the occurrence of such phenomenon.

It is also possible for the reconstruction to be induced by the presence of adsorbates on the surface. The fcc(110) surface could reconstruct into either pairing-row or missing-row structures [65–67]. For such reconstructions, the chemisorption of the adsorbate is known to be the driving force for the reconstruction. An example of this is the oxygen and alkali metal induced missing-row reconstruction of Pd [68, 69]. Interestingly, previous studies also found that this phenomenon also happens even for the case of adsorbed H [70, 71].

At the low coverage (e.g. 0.25 ML) regime, the adsorption of H was found to be stable at the pseudo-threefold hollow site of the Pd(110) surface. Hydrogen atoms remain on the pseudo-threefold site at 1 ML coverage. However, H atoms will favor a zigzag chain configuration ((2 × 1) structure) at such coverage (Fig. 2.29a). The

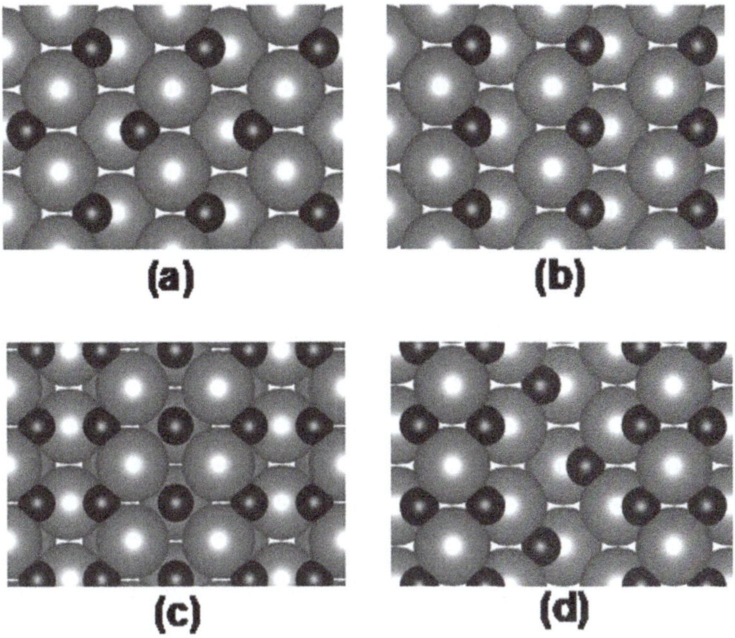

Fig. 2.29 1 ML coverage of H on unreconstructed Pd(110) surface: **a** (2 × 1) (zigzag chain) and **b** (1 × 1) (long chain) structures. 1.5 ML coverage of H on reconstructed Pd(110) surface: **c** pairing-row and **d** missing-row structures

adsorption energy per H atom in such case is −0.50 eV, as reported by Ledentu et al. [71]. The straight chain arrangement ((1 × 1) structure) (Fig. 2.30b) is also energetically stable but is less stable by 29 meV with respect to the zigzag chain system. The (2 × 1) structure provides a configuration such that the H atoms are located farther from each other than in the case of the (1 × 1) structure. Since the H atoms on metal surfaces tend to gain charge from the metal atoms, there will

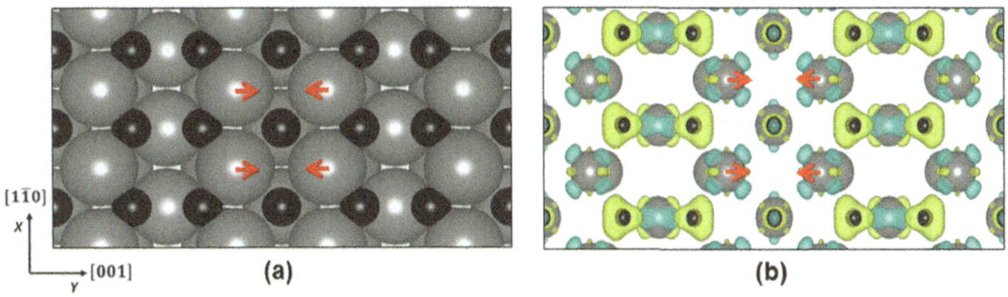

Fig. 2.30 **a** Geometric structure of the H-induced Pd(110) (1 × 2) pairing-row reconstructed surface for H coverage of 1.5 ML. **b** The charge density difference distribution of the adsorbed H atoms is also shown to illustrate that the repulsion among the adsorbed H atoms drive the reconstruction of the surface. Yellow and blue regions represent gain and loss of charge, respectively

be more repulsion between adsorbed H atoms in the (1 × 1) structure than in the (2 × 1) structure. Furthermore, it is expected that the Pd atoms will have better screening of the possible electrostatic interaction between the H atoms due to the zigzag arrangement in the (2 × 1) structure.

At a higher coverage of 1.5 ML, there would be a restructuring of the Pd(110) surface, in which a pairing-row (Fig. 2.29c) or missing-row (Fig. 2.29d) reconstruction could be realized. The pairing-row reconstructions readily happen when there is a large amount of H on the surface of Pd(110). It is induced by the repulsion of adsorbed H atoms (Fig. 2.30). On the surface, the surface atoms transfer charge to the adsorbed H atoms, which leads to the repulsion among the adsorbates. Since the (110) facet has an open structure (not compact), the Pd atoms are then pushed closer to each other, forming the pairing-row structure. The pairing-row reconstruction could be regarded as a metastable state due to the stronger adsorption energy of H on the missing-row reconstruction. This is confirmed based on the calculated average adsorption energies of H on the reconstructed surfaces. On pairing-row and missing-row structures, the adsorption energies per H atom at a coverage of 1.5 ML are −0.46 eV and −0.53 eV, respectively. These values suggest that the adsorbed H atoms will prefer to reside on the missing-row reconstructed surface. Furthermore, the stronger adsorption on such surface is also evident from its adsorption on the fcc hollow adsorption sites, which became accessible due to the reconstruction. Note that on Pd(111), H is most stable on the fcc hollow site. In an experiment, the pairing-row reconstruction is observed to occur only at low temperatures (below 200 K). Further exposure of H or increase in temperature will result to the restructuring of the system into the missing-row configuration. It can be inferred based from experimental findings and theoretical conclusions that in an actual operation of the the H–Pd(110) system, the missing-row configuration is the dominant structure while the pairing-row configuration is the metastable one.

The reconstruction of Pd(110) into a missing-row configuration is an activated process. This could be deduced from the fact that it happens at high H coverage, when there is enough H to induce the reconstruction. In addition, the reconstruction involves the migration of the surface atoms, which usually requires a certain amount of energy. Experimentally, the reconstruction is also observed even at low H coverage. However, such phenomenon happens only if it is at elevated temperatures (e.g. room temperature as shown in Fig. 2.31) [72, 73]. Such finding further signifies the activated process that exists in the formation of the adsorbate-induced missing-row reconstruction.

In a previous study, the mechanism of H-induced missing-row reconstruction was elucidated with focus on the initial stage of reconstruction [56]. DFT based calculations were performed to determine the energy profile of the diffusion of an adsorbed H atom and the simultaneous migration of Pd atom to other surface sites. Thus, the initial state entails an H atom that is initially on the surface region of the unreconstructed Pd. The final state is an H atom that is positioned at the lower surface region (subsurface), and a Pd atom that migrated to other surface sites denoting the reconstruction of the surface. Two different paths were considered: (1) the H atom is initially on the pseudo-threefold site and moves to the octahedral site in the

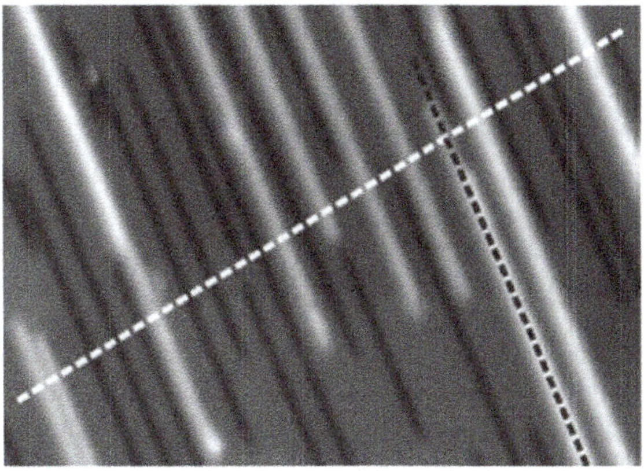

Fig. 2.31 STM image of the missing row reconstructed Pd(110) obtained after hydrogen exposure (<1 ML) at room temperature. Reprinted with permission from [72]. Copyright 2006 by Elsevier

subsurface, and (2) the H atom is initially on the long bridge site and moves to the tetrahedral site in the subsurface. The reaction paths, shown in Fig. 2.32, are obtained to assess the preferred pathway for H diffusion during the reconstruction process.

The obtained reaction path shows that the LB-TET configuration is more preferred over the P3F-OCT configuration due to a lower activation energy. However, the difference in the calculated activation energies is only 0.81 eV. Such value is small with reference to the obtained activation barriers of ~1.0 eV. Thus, it is not conclusive that the LB-TET configuration is the favored pathway for the adsorbate-induced reconstruction. In this regard, the H-induced reconstruction energy (E_{rec}) is proposed to further evaluate the energetics of the systems. It is defined by the equation:

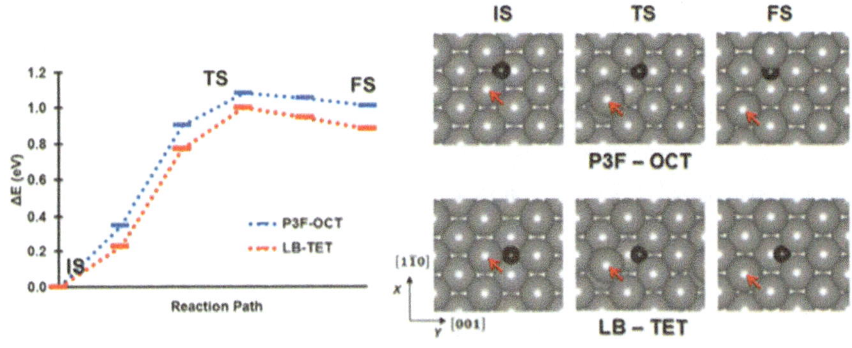

Fig. 2.32 Reaction paths of the H-induced missing-row reconstruction of the Pd(110). Two different paths were considered: the H atom is initially on the pseudo-threefold site and moves to the octahedral site in the subsurface (P3F-OCT), and the H atom is initially on the long bridge site and moves to the tetrahedral site in the subsurface (LB-TET). The Pd atom that migrates as H diffuses is marked by an arrow

$$E_{rec} = \left(E_{H/Pd-recon} - E_{Pd-recon} - \frac{1}{2}E_{H_2}\right) - \left(E_{H/Pd-unrecon} - E_{Pd-unrecon} - \frac{1}{2}E_{H_2}\right). \quad (2.9)$$

This will lead to

$$E_{rec} = \left(E_{H/Pd-recon} - E_{H/Pd-unrecon}\right) - \left(E_{Pd-recon} - E_{Pd-unrecon}\right). \quad (2.10)$$

The total energies of the H atoms adsorbed on the reconstructed and unreconstructed Pd(110) are defined by $E_{H/Pd-recon}$ and $E_{H/Pd-unrecon}$, respectively. The total energies of the reconstructed and unreconstructed Pd(110) surfaces are defined by $E_{Pd-recon}$ and $E_{Pd-unrecon}$, respectively. $E_{H/Pd-unrecon}$ is therefore the total energy of the H atom adsorbed on the P3F or LB site of the unreconstructed surface. In other words, E_{rec} defines the preference of the system to undergo reconstruction with or without the presence of H. A positive (negative) value indicates that the surface favors the reconstruction without (with) the H atom.

Figure 2.33 shows the plot of E_{rec} with respect to the obtained reaction path configurations. It shows that the initial step of reconstruction for both paths, the LB-TET and P3F-OCT configurations, favors the presence of adsorbate. This indicates the strong preference of the system to reconstruct in the presence of H. It further shows that the more energetically favored H-induced reconstruction is the LB-TET rather than the P3F-OCT configuration. It is therefore possible that the H atom takes the LB-TET pathway during the early stage of the reconstruction process. The geometric structure of the H–Pd system in the LB-TET configuration also supports the claim for this configuration. Along such path, H is always in a threefold configuration with respect to its interaction with the Pd atoms. Such configuration is the most favored adsorption structure of H on compact Pd surfaces. For the P3F-OCT configuration, only the initial state corresponds to a threefold coordination for the H atom, while the final state is a twofold coordinated. From these results, it can be concluded that

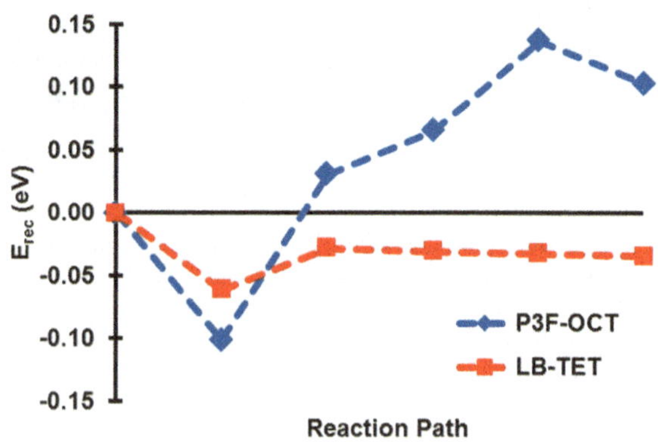

Fig. 2.33 Hydrogen-induced reconstruction energies for the pseudo-threefold to octahedral (P3F-OCT) and long bridge to tetrahedral (LB-TET) diffusion of the H atom

the interaction of H with the Pd atoms strongly affects the reconstruction process of the surface.

The fundamental concepts on the interaction of hydrogen and hydrogen-containing molecules with surfaces discussed in this Chapter have been laid to serve as a foundation for more complex surface phenomena. While DFT-based computational modeling of these reactions has proven to be indispensable in the analysis and design of materials with the desired selectivites and activities, the issues of accuracy, reliability, and corroboration with experiments remain to be some of the exciting fields of research in surface science.

References

1. H. Kasai, A. Okiji, J. Phys. Soc. Jpn. **50**, 2363 (1981)
2. H. Kasai, W.A. Diño, R. Muhida, Prog. Surf. Sci. **72**, 53 (2003)
3. B. Hammer, J.K. Nørskov, Surf. Sci. **343**, 211 (1995)
4. B. Hammer, Nature **376**, 238 (1995)
5. A.A.B. Padama, M.C.E. Escaño, H. Kasai, M.Y. David, H. Kawai, e-J. Surf. Sci. Nanotechnol. **8**, 3 (2010)
6. A.A.B. Padama, H. Kasai, H. Kawai, Surf. Sci. **606**, 62 (2012)
7. A. Groß, Surf. Sci. **606**, 690 (2012)
8. L. Kristinsdóttir, E. Skúlason, Surf. Sci. **606**, 1400 (2012)
9. S. Ogura, M. Okada, K. Fukutani, J. Phys. Chem. C **117**, 9366 (2013)
10. J.K. Nørskov, A. Houmøller, P.K. Johansson, B.I. Lundqvist, Phys. Rev. Lett. **46**, 257 (1981)
11. P. Ferrin, S. Kandoi, A.U. Nilekar, M. Mavrikakis, Surf. Sci. **606**, 679 (2012)
12. W. Dong, V. Ledentu, Ph. Sautet, A. Eichler, J. Hafner, Surf. Sci. **411**, 123 (1998)
13. J.K. Hammer, Nørskov. Adv. Catal. **45**, 71 (2000)
14. J.K. Nørskov, F. Studt, F. Abild-Pedersen, T. Bligaard, *Fundamental Concepts in Heterogeneous Catalysis* (Wiley, 2014)
15. M.J. Kolb, F. Calle-Vallejo, L.B.F. Juurlink, M.T.M. Koper, J. Chem. Phys. **140**, 134708 (2014)
16. Q. Fu, Y. Luo, ACS Catal. **3**, 1245 (2013)
17. D.A. McCormack, R.A. Olsen, E.J. Baerends, Chem. Phys. **122**, 194708 (2005)
18. Ž Šljivančanin, B. Hammer, Phys. Rev. B **65**, 085414 (2002)
19. G. Füchsel, K. Cao, S. Er, E.W. Smeets, A.W. Kleyn, L.B. Juurlink, G.J. Kroes, J. Phys. Chem. Lett. **9**, 170 (2017)
20. L. Árnadóttir, E.M. Stuve, H. Jónsson, Surf. Sci. **604**, 1978 (2010)
21. A. Michaelides, V.A. Ranea, P.L. De Andres, D.A. King, Phys. Rev. Lett. **90**, 216102 (2003)
22. S. Meng, E.G. Wang, S. Gao, Phys. Rev. B **69**, 195404 (2004)
23. A. Hodgson, S. Haq, Surf. Sci. Rep. **64**, 381 (2009)
24. J. Carrasco, B. Santra, J. Klimes, A. Michaelides, Phys. Rev. B **106**, 026101 (2011)
25. R.L. Arevalo, S.M. Aspera, M.C.S. Escano, H. Nakanishi, H. Kasai, ACS Omega **2**, 1295 (2017)
26. R.L. Arevalo, S.M. Aspera, M.C.S. Escano, H. Nakanishi, H. Kasai, Sci. Rep. **7**, 13963 (2017)
27. R.L. Arevalo, S.M. Aspera, M.C.S. Escano, H. Nakanishi, H. Kasai, J. Phys.: Condens. Matter **29**, 184001 (2017)
28. P.W. van Grootel, R.A. van Santen, E.J. Hensen, J. Phys. Chem. C **115**, 13027 (2011)
29. B.S. Bunnik, G.J. Kramer, J. Catal. **242**, 309 (2006)
30. G. Gajewski, C.W. Pao, J. Chem. Phys. **135**, 064707 (2011)
31. M.S. Liao, C.T. Au, C. F. Ng. Chem. Phys. Lett. **272**, 445 (1997)
32. A. Kokalj N. Bonini C. Sbraccia S. de Gironcoli S. Baroni, J. Am. Chem. Soc. **126**, 16732 (2004)

33. D. Ciuparu, M.R. Lyubovsky, E. Altman, L.D. Pfefferle, A. Datye, Catal. Rev. **44**, 593 (2002)
34. S.K. Matam, M.H. Aguirre, A. Weidenkaff, D. Ferri, J. Phys. Chem. C **114**, 9439 (2010)
35. R. Westerstrom, M.E. Messing, S. Blomberg, A. Hellman, H. Gronbeck, J. Gustafson, N.M. Martin, O. Balmes, R. van Rijn, J.N. Andersen, K. Deppert, H. Bluhm, Z. Liu, E. Grass, M. Havecker, E. Lundgren, Phys. Rev. B **83**, 115440 (2011)
36. R. van Rijn, O. Balmes, A. Resta, D. Wermeille, R. Westerstrom, J. Gustafson, R. Felici, E. Lundgren, J.W.M. Frenken, Phys. Chem. Chem. Phys. **13**, 13167 (2011)
37. A. Hellman, A. Resta, N.M. Martin, J. Gustafson, A. Trinchero, P.-A. Carlsson, Phys. Chem. Lett. **3**, 678 (2011)
38. J.F. Weaver, C. Hakanoglu, A. Antony, A. Asthagiri, Chem. Soc. Rev. **43**, 7536 (2014)
39. J.G. McCarty, Catal. Today **26**, 283 (1995)
40. R.L. Arevalo, S.M. Aspera, H. Nakanishi, Catal. Sci. Technol. **9**, 232 (2019)
41. A.V. Ruban, H.L. Skriver, J.K. Nørskov, Phys. Rev. B **59**, 15990 (1999)
42. O.M. Løvvik, Surf. Sci. **583**, 100 (2005)
43. S. Helfensteyn, J. Luyten, L. Feyaerts, C. Creemers, Appl. Surf. Sci. **212**, 844 (2003)
44. A. Rochefort, M. Abon, P. Delichere, J.C. Bertolini, Surf. Sci. **294**, 43 (1993)
45. J. Greeley, M. Mavrikakis, Nature Mater. **3**, 810 (2004)
46. B. Hammer, J.K. Nørskov, *Chemisorption and Reactivity on Supported Clusters and Thin Films*, ed. by R.M. Lambert, G. Pacchioni (Kluwer Academic Publishers, Berlin, 1997)
47. C.A. Menning, J.G. Chen, J. Chem. Phys. **130**, 174709 (2009)
48. A.A.B. Padama, H. Kasai, Y.W. Budhi, Int. J. Hydrogen Energy **38**, 14715 (2013)
49. Y.W. Budhi, I. Noezar, F. Aldiansyah, P.V. Kemala, A.A.B. Padama, H. Kasai, Int. J. Hydrogen Energy **36**, 15372 (2011)
50. Y.W. Budhi, H. Rionaldo, A.A.B. Padama, H. Kasai, I. Noezar, Int. J. Hydrogen Energy **40**, 10081 (2015)
51. H. Amandusson, L.G. Ekedahl, H. Dannetun, J. Membrane Sci. **193**, 35 (2001)
52. S. Uemiya, T. Matsuda, E. Kikuchi, J. Membrane Sci. **56**, 315 (1991)
53. J. Okazaki, D.A.P. Tanaka, M.A.L. Tanco, Y. Waku, F. Mizukami, T.M. Suzuki, J. Membrane Sci. **282**, 370 (2006)
54. A.B. Padama, H. Kasai, Y.W. Budhi, N.B. Arboleda Jr., J. Phys. Soc. Jpn. **81**, 114705 (2012)
55. A.A.B. Padama, H. Nakanishi, H. Kasai, Appl. Surf. Sci. **359**, 687 (2015)
56. A.A.B. Padama, H. Kasai, J. Chem. Phys. **140**, 244707 (2014)
57. A.A.B. Padama, H. Kasai, J. Alloys Compd. **645**, S123 (2015)
58. P. Kamakoti, D.S. Sholl, J. Membrane Sci. **225**, 145 (2003)
59. H. Okuyama, W. Siga, N. Takagi, M. Nishijima, T. Aruga, Surf. Sci. **401**, 344 (1998)
60. A.A.B. Padama, A.P.S. Cristobal, J.D. Ocon, W.A. Diño, H. Kasai, J. Phys. Chem. C **121**, 17818 (2017)
61. M.A. Saqlain, A. Hussain, M. Siddiq, A.A. Leitão, RSC Adv. **5**, 47066 (2015)
62. E. Vignola, S.N. Steinmann, B.D. Vandegehuchte, D. Curulla, P. Sautet, J. Phys. Chem. C **120**, 26320 (2016)
63. A.J. McCue, J.A. Anderson, J. Catal. **329**, 538 (2015)
64. T. Gritsch, D. Coulman, R.J. Behm, G. Ertl, Surf. Sci. **257**, 297 (1991)
65. R.J. Behm, V. Penka, M.G. Cattania, K. Christmann, G. Ertl, J. Chem. Phys. **78**, 7486 (1983)
66. M.G. Cattania, V. Penka, R.J. Behm, K. Christmann, G. Ertl, Surf. Sci. **126**, 382 (1983)
67. J. Yoshinobu, H. Tanaka, M. Kawai, Phys. Rev. B **51**, 4529 (1995)
68. R.A. Bennett, S. Poulston, I.Z. Jones, M. Bowker, Surf. Sci. **401**, 72 (1998)
69. S. Titmuss, A. Wander, D.A. King, Chem. Rev. **96**, 1291 (1996)
70. G. Kleinle, M. Skottke, V. Penka, G. Ertl, R.J. Behm, W. Moritz, Surf. Sci. **189**, 177 (1987)
71. V. Ledentu, W. Dong, P. Sautet, Surf. Sci. **412**, 518 (1998)
72. M. Kralj, C. Becker, K. Wandelt, Surf. Sci. **600**, 4113 (2006)
73. E. Kampshoff, N. Waelchli, A. Menck, K. Kern, Surf. Sci. **360**, 55 (2006)

Chapter 3
Quantum States and Dynamics of Hydrogen

Abstract Although the classical description of the atomic nuclei is sufficient in most scenarios, certain phenomena require a quantum mechanical treatment to fully grasp the nature of such particles. Since the quantum mechanical features are prominently associated with particles with small masses, hydrogen can exhibit quantum mechanical-based phenomena at relatively high temperatures as compared to other elements. This chapter will discuss various phenomena and applications where the quantum mechanical description of nuclei is vital. It starts with a brief review of the Born-Oppenheimer approximation, a necessary framework to treat the nuclei motions at their electronic ground state. The next two sections then discuss the hydrogen adsorption and hydrogen reaction dynamics on surfaces under the quantum mechanical framework, respectively. The chapter then concludes with a brief review of the measurement methods for hydrogen dynamics on surfaces.

3.1 The Born-Oppenheimer Approximation

Since the mass of an electron is lighter than a hydrogen nucleus by three orders of magnitude, the motion of the electron can be decoupled from the hydrogen nucleus, as well as the motion of hydrogen nucleus from the nuclei of surface atoms. This formalism, known as the Born-Oppenheimer approximation, allows the quantum mechanical treatment of both electrons and hydrogen nucleus, while the nuclei of surface atoms can be treated classically. This approximation was mentioned in Chap. 1 and is revisited in this Chapter.

Consider a molecule with N nuclei and n electrons. The Hamiltonian can be written as:

$$H = K_N + K_e + V_{NN} + V_{Ne} + V_{ee} \tag{3.1}$$

© Springer Nature Singapore Pte Ltd. 2020

H. Kasai et al., *Hydrogen and Hydrogen-Containing Molecules on Metal Surfaces*, Springer Series in Surface Sciences 71,
https://doi.org/10.1007/978-981-15-6994-4_3

where K_N, K_e, V_{NN}, V_{Ne}, and V_{ee} are the total kinetic energy of nuclei, the total kinetic energy of the electrons, the total potential energy of all nucleus-nucleus interactions, the total potential energy of all nucleus-electron interactions, and the total potential energy of all electron-electron interactions, respectively.

The kinetic energy terms in (3.1) in Cartesian coordinate take the form:

$$H = -\sum_{j=1}^{N} \frac{\hbar^2}{2M_j}\nabla^2_{\boldsymbol{R}_j} - \sum_{i=1}^{n} \frac{\hbar^2}{2m}\nabla^2_{\boldsymbol{r}_i} + V_{NN}(\boldsymbol{R}) + V_{Ne}(\boldsymbol{r}, \boldsymbol{R}) + V_{ee}(\boldsymbol{r}). \quad (3.2)$$

The variable $\boldsymbol{R} = (\boldsymbol{R}_1, \ldots, \boldsymbol{R}_N)$ represents the positions of the N nuclei, and $\boldsymbol{r} = (\boldsymbol{r}_1, \ldots, \boldsymbol{r}_N)$ represents the positions of the n electrons. For the sake of brevity, the potential terms V_{NN}, V_{Ne}, and V_{ee} are not written in their explicit forms, which essentially give the Coulomb interactions between particles.

By invoking the Born-Oppenheimer approximation, the electrons always adjust themselves to the motion of the nuclei adiabatically and are thereby in their eigenstates k. This assumption allows the decoupling of the electron motion from the nuclei motion. Consequently, the relevant Hamiltonian for the electrons has the form:

$$H_{Electron} = -\sum_{i=1}^{n} \frac{\hbar^2}{2m}\nabla^2_{\boldsymbol{r}_i} + V_{Ne}(\boldsymbol{r}, \boldsymbol{R}) + V_{ee}(\boldsymbol{r}). \quad (3.3)$$

From (3.3), the Schrödinger equation for the electrons with the internuclear potentials is:

$$\left[-\sum_{i=1}^{n} \frac{\hbar^2}{2m}\nabla^2_{\boldsymbol{r}_i} + V_{Ne}(\boldsymbol{r}, \boldsymbol{R}) + V_{ee}(\boldsymbol{r}) + V_{NN}(\boldsymbol{R})\right]\psi_k(\boldsymbol{r}; \boldsymbol{R}) = U_k(\boldsymbol{R})\psi_k(\boldsymbol{r}; \boldsymbol{R}). \quad (3.4)$$

The $\psi_k(\boldsymbol{r}; \boldsymbol{R})$ is the wavefunction of the electrons in the eigenstate k. The wavefunction for the electrons is a function of electron positions $\boldsymbol{r}$, while the nuclei positions $\boldsymbol{R}$ are treated as parameter (indicated by putting $\boldsymbol{R}$ after the semicolon). With nuclei motion and electrons motion decoupled, the ansatz of the total wavefunction as the sum of the nuclei wavefunction in their eigenstate K for a given electron eigenstate k, $\Phi_{K;k}(\boldsymbol{R})$, and multiplied by the wavefunction of the electrons can be written as follows:

$$\Psi_{k,K}(\boldsymbol{r}, \boldsymbol{R}) = \sum_{k,K} \Phi_{K;k}(\boldsymbol{R})\psi_k(\boldsymbol{r}; \boldsymbol{R}). \quad (3.5)$$

With the assumption that the electrons are in their ground state, the index k can be chosen so that $\psi_k(\boldsymbol{r}; \boldsymbol{R})$ is in the ground state, and then omitted for the sake of brevity. Any electronic excitations can then be further treated by other methods such as perturbation theory. The wavefunction then takes the form:

$$\Psi_K(\boldsymbol{r}, \boldsymbol{R}) = \Phi_K(\boldsymbol{R})\psi(\boldsymbol{r}; \boldsymbol{R}). \tag{3.6}$$

Considering the treatment of Berry's phase, the effective Hamiltonian for the nuclei takes the form [56]:

$$H_{Nuclear} = -\sum_{j=1}^{N} \frac{\hbar^2}{2M_j} \nabla^2_{\boldsymbol{R}_j} - \sum_{j=1}^{N} \frac{\hbar^2}{2M_j} \int dr \left|\nabla_{R_j} \phi_K(\boldsymbol{R})\right|^2 + U(\boldsymbol{R}) \tag{3.7}$$

where U is the energy of the electrons in their ground state as defined in (3.4). As a first approximation, the second term of (3.7) is usually omitted. The result is the lowest-order Born-Oppenheimer nuclear Hamiltonian:

$$H_{Nuclear} \cong -\sum_{j=1}^{N} \frac{\hbar^2}{2M_j} \nabla^2_{\boldsymbol{R}_j} + U(\boldsymbol{R}). \tag{3.8}$$

In practice, $U(\boldsymbol{R})$ can be obtained by any electronic structure calculation techniques such as first-principles calculations discussed in Chap. 1. The general procedure is to calculate $U(\boldsymbol{R})$ for a set of appropriate values of nuclei coordinate $\boldsymbol{R}$, and then calculate the hydrogen nuclei motions using the Hamiltonian in (3.8). Therefore, (3.8) and its reformulations will be necessary in describing the hydrogen quantum states throughout this Chapter.

3.2 Quantum States of the Hydrogen Adsorption on Solid Surfaces

This Section will discuss the quantum states of hydrogen atoms and molecules adsorbed on the surfaces. It will start with the simplest case of nucleus quantum states of hydrogen adsorption. Then, the more complicated case of hydrogen hindered rotation on the solid surfaces will be considered.

3.2.1 Hydrogen Atom Adsorption on Solid Surfaces

Consider the case of a low coverage dissociative hydrogen molecule adsorption on surfaces. In this scenario, the overlapping of hydrogen nuclei wavefunctions among adsorbed hydrogen atoms is negligible, which simplifies the problem to consider only a surface with one hydrogen atom. The Hamiltonian has the form:

$$H_{Nuclear} = -\sum_{j=1}^{N} \frac{\hbar^2}{2M_j} \nabla^2_{\boldsymbol{R}_j} + U(\boldsymbol{R}). \tag{3.9}$$

$U(\boldsymbol{R})$ is then calculated for a set of appropriate values of nuclei coordinate $\boldsymbol{R}$ as described in Sect. 3.1, which finally gives the adiabatic potential that can be used to calculate the nuclear wavefunctions. Variation in hydrogen mass is also treated in the potential energy $U(\boldsymbol{R})$, and thus, the isotope effect can be analyzed as well.

As an example, consider the nuclear wavefunctions of hydrogen atoms adsorbed on the (111) surfaces of face-centered cubic (fcc) structures of Pt and Cu. The structure of an fcc (111) surface is shown in Fig. 3.1. To construct $U(\boldsymbol{R})$, a grid composed of different values of $\boldsymbol{R}$ is constructed, as shown in the bottom left of Fig. 3.1. The next step is to calculate the potential energy of H atom at each point in the grid, with varying heights of H from the surface, to obtain $U(\boldsymbol{R})$. The results for Cu(111) and Pt(111) are presented in Fig. 3.2a and b, respectively.

The Potential Energy Curves (PECs) in Fig. 3.2 show that the depths of the potential wells of H atom on Cu(111) vary as H atom moves parallel the surface plane. On the contrary, the depths are nearly identical in the case of H atom on Pt(111). From the PEC profiles, it is expected that the H motion on Cu(111) will be relatively localized as compared to its motion on Pt(111). To calculate the wavefunctions, it is needed to obtain a detailed or complete set (by taking many points in the grid in Fig. 3.1) of PECs. However, performing first-principles calculations of such PECs is computationally expensive, and it is more practical to perform interpolation from the PECs obtained using few points in the grid. In this example, the interpolations were performed by fitting parameters into the Morse potential:

$$U(x, y) = D(x, y)\left[\left\{e^{-\alpha(x,y)\times(z-z_{ad}(x,y))} - 1\right\}^2 - 1\right]. \tag{3.10}$$

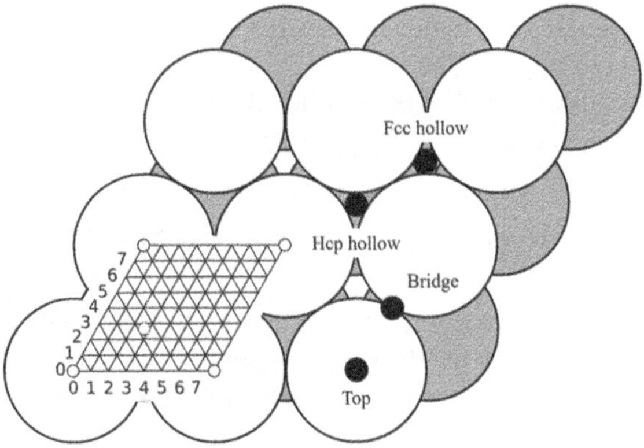

Fig. 3.1 Atomic structure of Cu(111) and Pt(111) viewed from the top view, perpendicular to the surface. White and gray circles represent atoms in the first and second layers, respectively. Black circles indicate on-top, bridge, fcc hollow, and hcp hollow. The lower left grid shows the chosen of coordinate points parallel to the surface. Adapted with permission from [55]. Copyright 2013 by Asakura Publishing

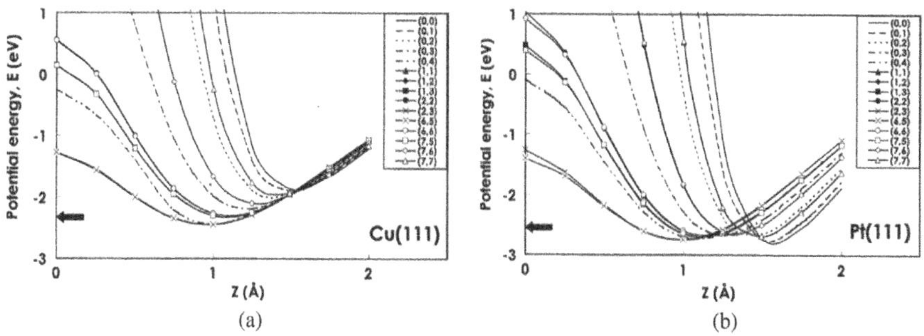

Fig. 3.2 Potential energy curves for hydrogen atom on Cu(111) and Pt(111) (Reproduced from the literature [1]). **a** The potential energy curves of H on Cu(111) for each coordinate point parallel to the surface plane. The horizontal axis (coordinate z) is the distance of H perpendicular to the surface, and the vertical axis is the potential energy. (0, 0) corresponds to the on-top site, and (2, 3) corresponds to the hcp hollow site. **b** The potential energy curves of H on Pt(111) [1, 2]. Reprinted with permission from [1]. Copyright 2001 by Elsevier

The optimal adsorption height, depth of the potential well, and width of the potential well correspond to $z_{ad}(x, y)$, $D(x, y)$, and $\alpha(x, y)^{-1}$, respectively. The results for Cu(111) and Pt(111) are shown in Figs. 3.3 and 3.4, respectively.

The depth $D(x, y)$ of the potential energy surfaces (PESs) shown in Figs. 3.3a and 3.4a predict the preferred adsorption sites of H atoms that are treated classically. Figs. 3.3 and 3.4 show that the local minima of the PES on Cu(111) is located only at the hollow sites (Fig. 3.3a). On the other hand, the local minima of the PES on Pt(111) are located on both the top sites and the hollow sites (Fig. 3.4a), with the adsorption on the top site slightly stronger than that of the hollow site. Nonetheless, the distribution of the optimal adsorption heights z_{ad} on both Cu(111) and Pt(111) is almost identical, as shown in Figs. 3.3b and 3.4b, respectively.

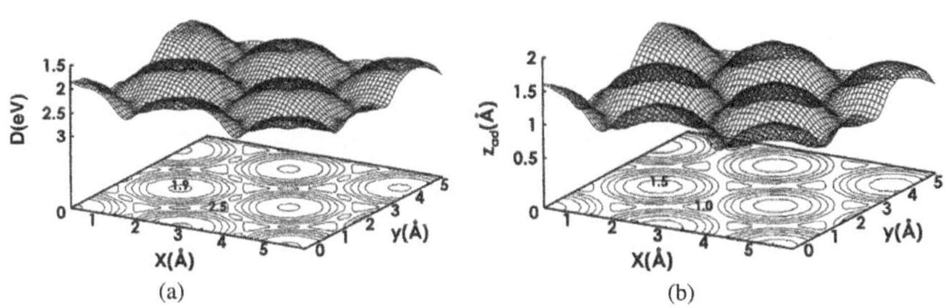

Fig. 3.3 **a** The depth $D(x, y)$ of the Morse potential well expressed as a function of real space coordinates (x, y) parallel to the surface (note that the higher the value, the weaker the potential energy). The contour interval is 0.05 to eV. **b** The optimal adsorption height z_{ad} of the potential well expressed as a function of the real space coordinate (x, y) parallel to the surface. Contour line interval is 0.1 Å. Reprinted with permission from [1]. Copyright 2001 by Elsevier

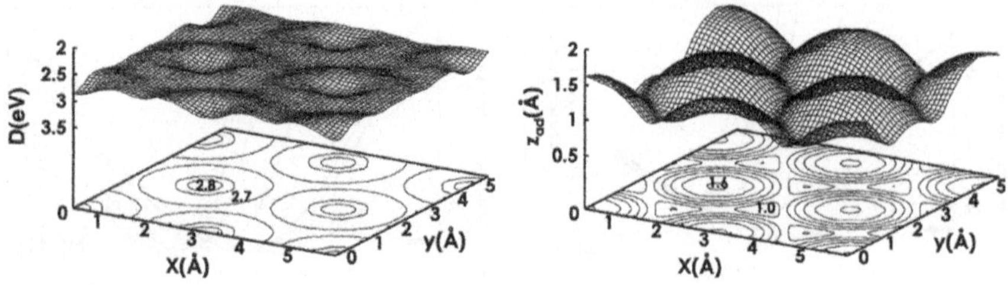

(a) (b)

Fig. 3.4 **a** The depth $D(x, y)$ of the Morse potential well and **b** the optimal adsorption height z_{ad} of the potential well for Pt(111). The notations are the same with Fig. 3.3. Reprinted with permission from [1]. Copyright 2001 by Elsevier

In the last step, the hydrogen nuclear wavefunctions are calculated numerically. In this example, the authors of Reference [1] used Gaussian type orbitals to construct the wavefunctions. The Gaussian type orbital is of the form:

$$\phi_i(x, y, z) = \left(\frac{\beta}{\pi}\right)^{3/4} e^{-\frac{\beta}{2}\left[(x-X_i)^2+(y-Y_i)^2+(z-Z_i)^2\right]} \tag{3.11}$$

where the index i labels the Gaussian-type orbital i whose center is positioned at (X_i, Y_i, Z_i), and β is adjusted to an adequate overlapping of the neighboring orbitals. $\beta = 102.7$ Å^{-2} (87.19 Å^{-2}) for the lattice constant of 3.62 Å (3.92 Å) corresponding to Cu(111) (Pt(111)). The ground-state nuclear wavefunctions of Cu(111) and Pt(111) are shown in Fig. 3.5a and b, respectively.

Results from Fig. 3.5a show that the nuclear wavefunctions of H adsorption on the Cu(111) is localized on the hollow site. On the other hand, the nuclear wavefunctions

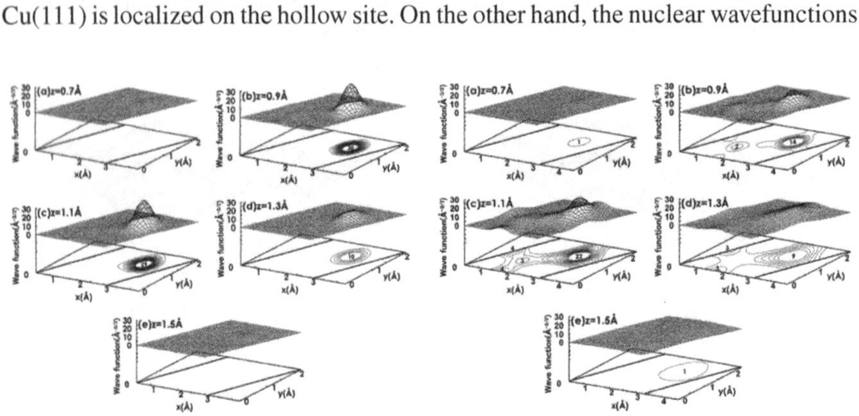

(a) (b)

Fig. 3.5 Ground state wavefunctions of hydrogen atom on **a** Cu(111) and **b** Pt(111) at different heights from the surface (positive values of z means the position is away from the surface towards the vacuum). The contour interval is 1.0 Å$^{-3/2}$. Reprinted with permission from [1]. Copyright 2001 by Elsevier

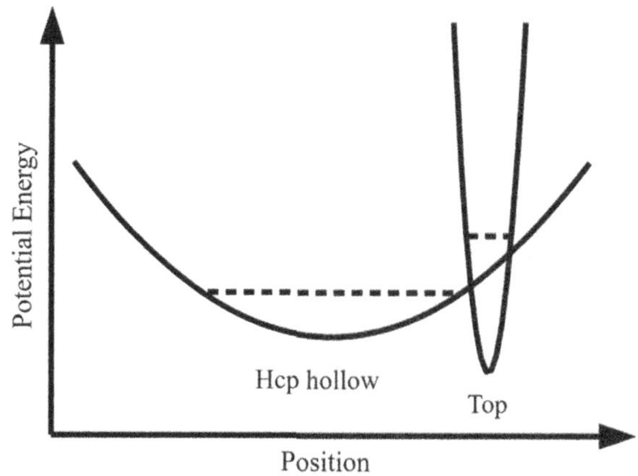

Fig. 3.6 Relationship between potential well and adsorption position considering quantum effects in Pt(111). The solid line shows potential energy wells corresponding to the top site and hcp hollow site. The broken line defines the zero-point vibrational levels. Adapted with permission from [55]. Copyright 2013 by Asakura Publishing

on Pt(111) is delocalized between the top and hollow sites, as shown in Fig. 3.5b. Another prominent result is that after including the nuclear wavefunctions, the most preferential adsorption sites on Pt(111) is changed from the top site to the hollow site. The reason is that the potential well for H atom adsorption on the top site is narrower than that of the hollow site. Consequently, the nuclear wavefunctions on the top site must have shorter wavelength than that of the hollow site. Since kinetic energy is larger for a particle with shorter wavelength in quantum mechanics, the H atom adsorbed on the top site have larger ground-state kinetic energy (the so called the zero-point energy) than the H atom adsorbed on the hollow site. The explanation is illustrated in Fig. 3.6.

3.2.2 *Hydrogen Molecule Hindered Rotation on Solid Surfaces*

This Section will discuss the hydrogen molecule that is weakly adsorbed on surfaces. This weakly adsorbed H_2 is an ideal situation to observe the ortho-para conversion phenomenon, which will be discussed in Sect. 3.2.3. In such case, the adsorbed H_2 can have rotational motion in addition to its vibrational motion.

3.2.2.1 Rotational States of the Isolated H_2

A briefly review of the two-body rotational states is first provided. For rotational motion, it is more appropriate to describe the motion using a coordinate with respect to the center-of-mass (CM), which can be done through the following transformation:

$$\boldsymbol{R}_{\mathrm{CM}} = \frac{1}{M}(M_1\boldsymbol{R}_1 + M_2\boldsymbol{R}_2), \tag{3.12}$$

$$\boldsymbol{r}_{\mathrm{CM}} = \boldsymbol{R}_1 - \boldsymbol{R}_2, \tag{3.13}$$

where $M = M_1 + M_2$ is the total nuclear mass, $|\boldsymbol{r}_{\mathrm{CM}}| = r_{\mathrm{CM}}$ is the nuclear distance of hydrogen molecule, and $\boldsymbol{R}_{\mathrm{CM}}$ is coordinate of the center-of-mass. Note that $\boldsymbol{r}_{\mathrm{CM}}$ is the center-of-mass coordinate for the nuclei of H_2 molecule and should not be confused with the coordinate for the electrons $\boldsymbol{r}$ as defined in the previous sections. Substituting (3.12) and (3.13) into (3.8), the nuclear motion equation becomes:

$$\left\{-\frac{\hbar^2}{2}\left[\frac{1}{M}\nabla^2_{\boldsymbol{R}_{\mathrm{CM}}} + \frac{1}{\mu}\nabla^2_{\boldsymbol{r}_{\mathrm{CM}}}\right] + U(\boldsymbol{R}_{\mathrm{CM}}, \boldsymbol{r}_{\mathrm{CM}})\right\}$$
$$\varphi_K(\boldsymbol{R}_{\mathrm{CM}}, \boldsymbol{r}_{\mathrm{CM}}) = E_K \varphi_K(\boldsymbol{R}_{\mathrm{CM}}, \boldsymbol{r}_{\mathrm{CM}}) \tag{3.14}$$

where μ is the reduced mass defined as $\frac{1}{\mu} = \frac{1}{M_1} + \frac{1}{M_2}$. The kinetic energy part of the Hamiltonian in (3.14) is composed of the center-of-mass kinetic energy and the internal kinetic energy. Since the interest is on H_2 molecule that rotates on the surface, the Laplacian operator $\nabla^2_{\boldsymbol{r}_{\mathrm{CM}}}$ of the internal kinetic energy, from the Cartesian coordinate to the spherical coordinate can be written as:

$$\nabla^2_{\boldsymbol{r}_{\mathrm{CM}}} = \frac{1}{r}\frac{\partial^2}{\partial r^2}r + \frac{1}{r^2}\left(\frac{1}{\sin\theta}\frac{\partial}{\partial\theta}\sin\theta\frac{\partial}{\partial\theta} + \frac{1}{\sin^2\theta}\frac{\partial^2}{\partial\phi^2}\right), \tag{3.15}$$

where the subscript CM in the radial coordinate r was omitted for the sake of brevity. Substituting (3.15) into (3.14) yields:

$$\left\{-\frac{\hbar^2}{2}\left[\frac{1}{M}\nabla^2_{\boldsymbol{R}_{\mathrm{CM}}} + \frac{1}{r}\frac{\partial^2}{\partial r^2}r\right] + \frac{1}{2\mu r^2}\boldsymbol{L}^2 + U(\boldsymbol{R}_{\mathrm{CM}}, r, \theta, \phi)\right\}$$
$$\varphi_K(\boldsymbol{R}_{\mathrm{CM}}, r, \theta, \phi) = E_K \varphi_K(\boldsymbol{R}_{\mathrm{CM}}, r, \theta, \phi), \tag{3.16}$$

where $\boldsymbol{L}^2$ is defined by:

$$\boldsymbol{L}^2 = -\hbar^2\left(\frac{1}{\sin\theta}\frac{\partial}{\partial\theta}\sin\theta\frac{\partial}{\partial\theta} + \frac{1}{\sin^2\theta}\frac{\partial^2}{\partial\phi^2}\right). \tag{3.17}$$

Here, $\boldsymbol{r}_{\mathrm{CM}}$ is written in spherical coordinates (r, θ, ϕ) and the subscript CM is omitted for the sake of brevity. The Hamiltonian can be further simplified by using the center-of-mass reference frame. In this reference frame, the kinetic energy of the center-of-mass vanishes:

$$\left\{-\frac{\hbar^2}{2}\left[\frac{1}{r}\frac{\partial^2}{\partial r^2}r\right] + \frac{1}{2\mu r^2}\boldsymbol{L}^2 + U(r, \theta, \phi)\right\}$$
$$\varphi_K(r, \theta, \phi) = E_K \varphi_K(r, \theta, \phi). \tag{3.18}$$

Assuming that the H_2 molecule is a rigid rotor by fixing $r = r_{eq}$, where r_{eq} is the equilibrium bond length, and if $U(r, \theta, \phi)$ only depends on the coordinate r as in the Coulomb potential, (3.18) can be separated into radial part and spherical part. The spherical part is given by:

$$-\frac{\hbar^2}{2\mu r_{eq}^2}\left(\frac{1}{\sin\theta}\frac{\partial}{\partial\theta}\sin\theta\frac{\partial}{\partial\theta}+\frac{1}{\sin^2\theta}\frac{\partial^2}{\partial\phi^2}\right)$$

$$\varphi_{J,m}(\theta, \phi) = E_{J,m}\varphi_{J,m}(\theta, \phi), \tag{3.19}$$

in which, the eigenenergy is:

$$E_{J,m} = \frac{\hbar^2}{2\mu r_{eq}^2}J(J+1), \tag{3.20}$$

and the eigenstates are the spherical harmonics

$$\varphi(\theta, \phi) = Y^{J,m}(\theta, \phi). \tag{3.21}$$

3.2.2.2 Correlation Between Nuclear Rotational States and Nuclear Spin States

In addition to the nuclear rotational states, the nuclear spin states are also important. The total nuclear wavefunction is given by

$$\Phi = \varphi_K(r, \theta, \phi)\chi(I), \tag{3.22}$$

where Φ is the total nuclear wavefunction from (3.6), φ_K is the spatial nuclear wavefunction from (3.14), and $\chi(I)$ is the nuclear spin wavefunction. Nuclear spins are unique to each isotope of each element. For example, light hydrogen nuclei are protons, which are fermions with spin $\frac{1}{2}$, while deuterium nuclei are bosons with spin 1. In the case of molecules composed of fermions, the indistinguishability of the particles demands that the spin and the orbital wave functions must be a combination of symmetric/antisymmetric or antisymmetric/symmetric wave functions. In the case of bosons, both must be symmetric or antisymmetric. Lastly, isomers generated by different nuclear spin states of molecules are called nuclear spin isomers.

As an example, consider the case of a molecule with two light hydrogen atoms that are isolated in vacuum. When the electronic system and the vibrational state are in the ground state, there only exists the angular momentum contributed by the nuclear rotational states and nuclear spin states. Since protons are fermions, the exchange in the positions of the nuclei $\varphi(\theta, \phi) = Y^{J,m}(\theta, \phi)$ must be antisymmetric:

$$Y^{J,m}(\pi-\theta, \pi+\phi) = (-1)^J Y^{J,m}(\theta, \phi). \tag{3.23}$$

On the other hand, the radial component of the spatial nuclear wavefunction is invariant to the exchange in the positions of nuclei:

$$R(r; \pi - \theta, \pi + \phi) = R(r; \theta, \phi). \tag{3.24}$$

Therefore, the symmetric and antisymmetric property of the nuclear wavefunction only appears in the angular part in (3.23). Even (odd) values of the angular momentum J correspond to (anti)symmetric $Y^{J,m}(\theta, \phi)$. Therefore, a rotational wave function for an odd number J corresponds to a symmetric (ortho) spin wave function

$$\chi_o(s, s') = \begin{cases} \sigma_\uparrow(s)\sigma_\uparrow(s') \\ \frac{1}{\sqrt{2}}\left[\sigma_\uparrow(s)\sigma_\downarrow(s') + \sigma_\downarrow(s)\sigma_\uparrow(s')\right] \\ \sigma_\downarrow(s)\sigma_\downarrow(s') \end{cases} \tag{3.25}$$

and an even number J for an antisymmetric (para) wave function

$$\chi_p(s, s') = \frac{1}{\sqrt{2}}\left[\sigma_\uparrow(s)\sigma_\downarrow(s') - \sigma_\downarrow(s)\sigma_\uparrow(s')\right], \tag{3.26}$$

where $\sigma_\uparrow(s)$ and $\sigma_\downarrow(s)$ are the nuclear spin functions in the up and down states at the spin coordinate s, respectively. A hydrogen molecule with a symmetric nuclear spin is called ortho hydrogen, and an antisymmetric one is called para hydrogen. Since the ortho-spin wave function $\chi_o(s, s')$ is triply degenerate, the abundance ratio of ortho-hydrogen and para-hydrogen at room temperature is 3:1. Since the excitation energy from $J = 0$ to 1 is 14.7 meV, the proportion of parahydrogen at $j = 0$ increases when thermal equilibrium is reached at a temperature of 170 K or lower. At the liquefaction temperature (20.55 K), the equilibrium concentration of parahydrogen is 99.8%.

3.2.2.3 Rotational States of H_2 Adsorbed on the Solid Surfaces

The full Hamiltonian of the H_2 adsorbed on the surfaces has similar form with the isolated H_2 in (3.14), except that the potential $U(\boldsymbol{R}_{CM}, \boldsymbol{r}_{CM})$ must take into account the influence of the surface. Consider the full model Hamiltonian of the H_2 molecule adsorbed on the surface, in which the Hamiltonian takes the form:

$$\left\{-\frac{\hbar^2}{2}\left[\frac{1}{M}\nabla^2_{\boldsymbol{R}_{CM}} + \frac{1}{\mu}\nabla^2_{\boldsymbol{r}_{CM}}\right] + U(\boldsymbol{R}_{CM}, \boldsymbol{r}_{CM})\right\}$$
$$\varphi_K(\boldsymbol{R}_{CM}, \boldsymbol{r}_{CM}) = E_K \varphi_K(\boldsymbol{R}_{CM}, \boldsymbol{r}_{CM}). \tag{3.27}$$

Since we are interested in the rotational states of H_2 adsorbed on the surface, it is preferable to change the internal coordinates ($\boldsymbol{r}_{CM}$) from Cartesian to spherical coordinates $\boldsymbol{r}_{CM} \to (r, \theta, \phi)$. For the sake of convenience which will be made clear to the reader, the coordinates $\boldsymbol{R}_{\mathrm{CM}} \rightarrow (X, Y, Z)$ will also be changed. The H_2 molecule adsorbed on the surface with these new set of coordinates are depicted in Fig. 3.7. With these new set of coordinates, and after dropping subscripts CM for the sake of brevity, the potential U then takes the form:

$$U(\boldsymbol{R}_{\mathrm{CM}}, \boldsymbol{r}_{\mathrm{CM}}) = U(\boldsymbol{R}_{\mathrm{CM}}, r, \theta, \phi) = U(X, Y, Z, r, \theta, \phi) \tag{3.28}$$

Consider the hindered rotation in the case of H_2 adsorption on the Pd(210) surface as an example. The Pd(210) with appropriate amount of preadsorbed hydrogen atom was recently found to be exceptional for ortho-para conversion than the conventional Ag(111) [3]. On this surface, the directions parallel to the surface are [001] and $[1\bar{2}0]$, as shown in Fig. 3.7.

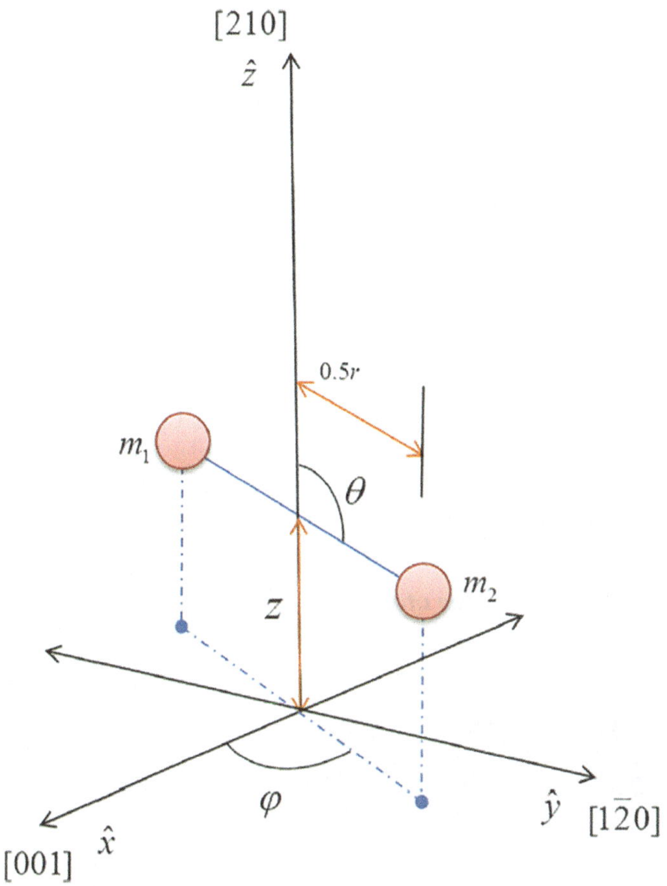

Fig. 3.7 The definitions of the coordinate of H_2 on the surface. The $X - Y$ plane is spanned in [001] and $[1\bar{2}0]$ directions. The parameters r and Z are the H_2 bond length and the height from center-of-mass of H_2 to the surface, respectively. The φ and θ are the azimuthal and the polar angles, respectively. Reprinted with permission from [4]. Copyright 2018 by the American Institute of Physics

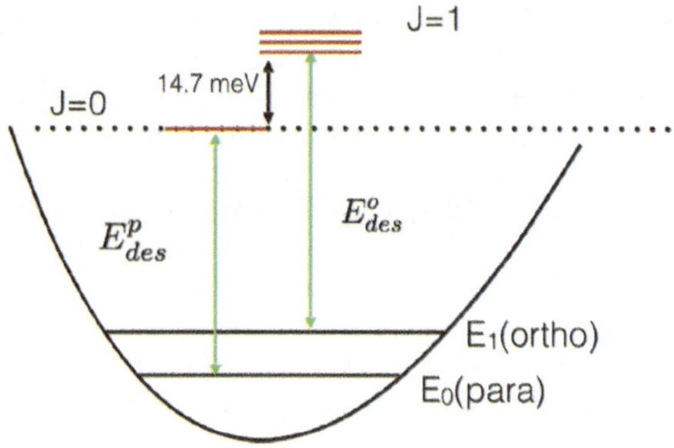

Fig. 3.8 Schematic diagram showing the desorption energies of the H_2 in the ortho and para nuclear spin isomers. The black curve is the potential energy well, and the dotted-black curve is the isolated para hydrogen molecule. Reprinted with permission from [3]. Copyright 2018 by the American Vacuum Society

The schematic for ortho and para H_2 adsorption on the surface can be depicted in the energy diagram as shown in Fig. 3.8. The energy difference between ortho and para H_2 in the gas phase is 14.7 eV. In the next section, the conversion between the ortho and para hydrogen on the catalyst surfaces will be discussed further.

The next step is to simplify the Hamiltonian in (3.27) as necessary. For a system at low temperatures, the lateral translational contributions of H_2 in the (X, Y) directions, as well as the interatomic distance r can be represented by their respective equilibrium values. With this approximation, (X, Y, r) can be dropped from the potential U, and the Hamiltonian has the form:

$$H = -\frac{\hbar^2}{2M}\frac{\partial^2}{\partial Z^2} + \frac{1}{2\mu r_{eq}^2}\boldsymbol{L}^2 + U(Z, \theta, \phi). \tag{3.29}$$

On the Pd(210) surface that is passivated by the pre-adsorbed hydrogen atom [4], the most preferable adsorption site for H_2 are the step edge sites as shown in Fig. 3.9 (blue color atoms). It was previously found that H_2 adsorbed on this particular state has constant U for any given value of ϕ [4], allowing to further simplify the potential

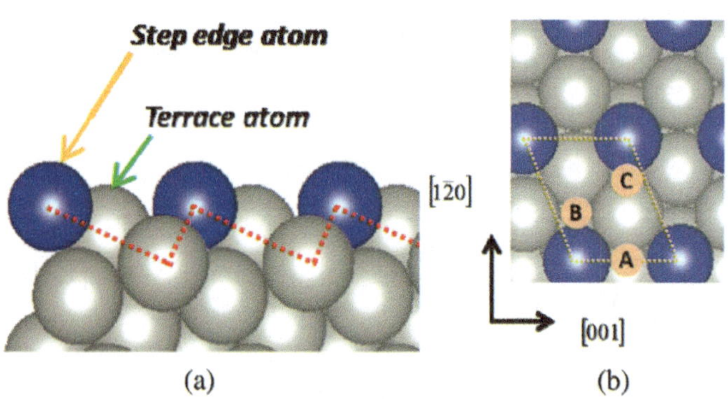

Fig. 3.9 The model of the Pd(210) surface. **a** The blue atoms are the step edge atoms. **b** The Labels A, B, C are the adsorption sites of the pre-adsorbed hydrogen atom. Reprinted with permission from [4]. Copyright 2018 by the American Institute of Physics

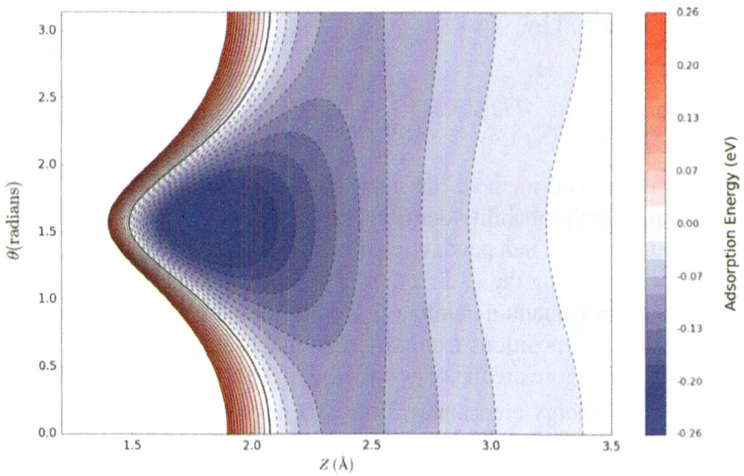

Fig. 3.10 The 2D adiabatic PES of H_2 on top of step-edge of Pd(210) with $\Theta = 1.0$ ML. The parameter Z is the molecule center-of-mass from the surface, and θ is the polar orientation. Reprinted with permission from [3]. Copyright 2018 by the American Vacuum Society

from 3D to 2D as follows:

$$H = -\frac{\hbar^2}{2M}\frac{\partial^2}{\partial Z^2} + \frac{1}{2\mu r_{eq}^2}\boldsymbol{L}^2 + U(Z,\theta). \tag{3.30}$$

The result of the PES $U(Z,\theta)$ from the first-principles calculations based on DFT is shown in Fig. 3.10. Similar to the hydrogen atom adsorption previously discussed, the authors of [3] also used Morse potential function to fit the PES as shown in Fig. 3.10.

The last step is to construct the wave functions. To numerically compute the wavefunction, in which, the following ansatz is used:

$$\Psi(Z,\theta,\phi) = \sum_{J,m,n} c_{J,m,n} e^{im\phi} \Phi_{J,m,n}(Z,\theta), \tag{3.31}$$

where the first term, $c_{J,m,n}$, is the coefficient. The second term, $e^{im\phi}$, is the solution of the wavefunction for azimuthal coordinate. It is a planewave since the potential $U(Z,\theta)$ is constant regardless of the value of ϕ. The last term is the Guassian function $G_n(Z)$ multiplied by the associated Legendre polynomials $P_J^m(\cos\theta)$ and the appropriate normalization factor:

$$\Phi_{Jmn}(Z,\theta) = G_n(Z)\sqrt{\frac{(2J+1)(J-m)!}{4\pi(J+m)!}}P_J^m(\cos\theta), \tag{3.32}$$

where $G_n(Z)$ is defined by

$$G_n(Z) = \left(\frac{\beta}{\pi}\right)^{1/4} e^{-\beta(Z-z_n)^2}. \tag{3.33}$$

Using the variational method, the nuclear wavefunction is solved for different eigenstates, and their probability densities $|\Psi(Z,\theta)|^2$ are shown in Fig. 3.11. The energy depends on three independent parameters shown as $E(\nu, J, m)$, where ν is the vibrational state along the Z direction, J is the total angular quantum number, and m is the magnetic quantum number. It is noted that the wavefunctions shown in Fig. 3.11 require superpositions from different eigenstates of ν, J, m, and thus the values shown in the figure are the ones with the highest contributions (largest value of $|c_{J,m,n}|^2$) to the energy eigenstates. With this notation, $E(0,0,0)$ is the ground state (Fig. 3.11a), which is a para hydrogen. The first excited state is the ortho

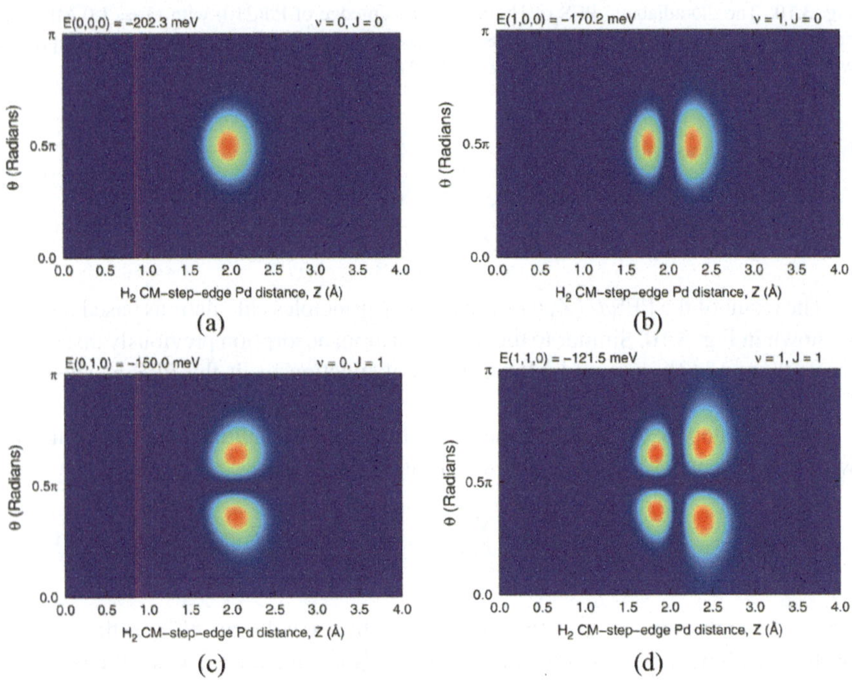

Fig. 3.11 The probability density $|\Psi(Z,\theta)|^2$ of nuclear wavefunctions of para H_2 adsorbed on the step edge of Pd(210). The plots adopted the symbol $E(\nu, J, m)$ to indicate the states of the nuclear wavefunctions, where ν is the vibrational state along the Z direction, J is the total angular quantum number, and m is the magnetic quantum number. In contrast to the H_2 in the gas phase where ν, J, m are the good quantum numbers, H_2 adsorbed on the surface requires superpositions from different values of ν, J, m; their values shown in the figures are the *highest* contributions (largest value of $|c_{J,m,n}|^2$) to the energy eigenstates. Reprinted with permission from [3]. Copyright 2018 by the American Vacuum Society

hydrogen with the notation $E(0, 1, 1)$ corresponding to the energy -197.6 meV. It has identical nuclear wavefunction $|\Psi(Z, \theta)|^2$ with that of $E(0, 0, 0)$ (its wavefunction is not shown in Fig. 3.11). When H_2 desorbs from the surface, the desorbed molecule assumes its corresponding ortho or para hydrogen in the gas phase, as shown schematically in Fig. 3.8. The desorption energy of ground state hydrogen $E(0, 0, 0)$ is estimated to be 202.3 meV, while the desorption energy of the first excited state hydrogen $E(0, 1, 1)$ is estimated to be 197.6 meV + 14.7 meV = 212.3 meV. That is, the difference between the ortho and para adsorption energy is $\Delta^{o-p} = 10$ meV. In comparison to the flat surfaces of Ni(111) and Cu(100), their Δ^{o-p} values are only 2 meV [5] and 1 meV [6], respectively. The large difference in the desorption energies between o–H_2 and p–H_2 on Pd(210) over the flat surfaces is very favorable for separating o–H_2 and p–H_2 isomers. Other examples of the different bound nuclear states are shown in Fig. 3.11b–d. Figure 3.11b, shows an example of the vibrationally excited wavefunction, Figure c shows an example of the rotationally excited wavefunction, and lastly Fig. 3.11d shows an example of both vibrationally and rotationally excited wavefunction. Note that all of the wavefunctions in Fig. 3.11 are all para hydrogen ($m = 0$; a nuclear spin with an even number).

3.2.3 Nuclear Spin Isomers Conversion of Hydrogen Molecule on Solid Surfaces

As discussed in Sect. 3.2.2.2, hydrogen molecules can have nuclear spin isomers, namely the para hydrogen (even value of J), and ortho hydrogen (odd value of J). The ratio between the ortho and para species is temperature dependent. In the gas phase, collisions between molecules lead to ortho-para conversion, but because it is a spin-forbidden transition, the conversion rate is very slow (about 3 years at room temperature and atmospheric pressure) [7]. However, the conversion can be accelerated when H_2 adsorbs on the appropriate surfaces.

Ortho-para conversion on the surface can occur in both chemisorption and physisorption systems. In chemisorption systems such as copper surfaces, ortho-para conversion can occur by indirect scattering via thermal equilibrium. In this process, the first step is the H_2 dissociative adsorption on the surface. After the molecule adsorbed and dissociated into H atoms, both atoms are no longer correlated to each other. The subsequent associative desorption of the atoms into molecules with arbitrary nuclear spin is possible, regardless of the initial conditions. On the other hand, in physical adsorption systems such as iron oxide and silver surfaces, ortho-para conversion by direct scattering can occur due to the interaction between the asymmetric surface magnetic field and the magnetic field associated with the nuclear spin of H_2. Since the mechanism of ortho-para conversion of the chemisorption system is self-explanatory, this section will discuss only the case of the physical adsorption system.

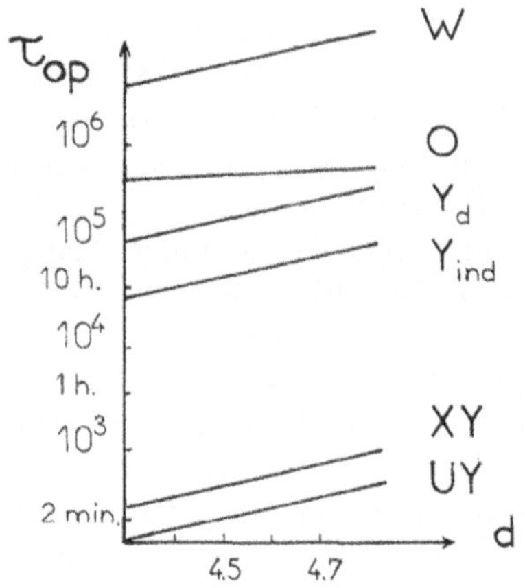

Fig. 3.12 The ortho-para conversion time τ_{op} on Ag(111) surface as a function of the metal-molecule distance d (in a.u.). The processes W, O, Y-$_d$ and Y_{ind} are the first-order processes by the magnetic dipole interaction, spin-orbit interaction, direct Fermi contact interaction, and indirect Fermi contact interaction, respectively. The processes XY and UY are the second-order processes. Reprinted with permission from [9]. Copyright 1992 by Elsevier

Various mechanisms have been proposed on the mechanism of ortho-para conversion based on the first-order and second-order perturbation theory. In the first-order perturbation theory, the process can be interpreted as the H_2 nuclear spin conversion without an intermediate step. On the other hand, in the second-order [8] (and higher-order) perturbation theory, the process signifies that there is (are) one (more than one) intermediate step(s) involve in the nuclear spin conversion process. Figure 3.12 shows the ortho-para conversion time for various first-order and second-order processes. The first-order processes W, O, Y_d and Y_{ind} are the magnetic dipole interaction, spin-orbit interaction, direct Fermi contact interaction, and indirect Fermi contact interaction, respectively. The second-order processes are the so-called XY and UY processes. Since the UY processes are the fastest, the following discussion will focus on these processes. The explanation for other processes can be found in other literatures such as [9–11].

Ortho-para conversion without nuclear exchange is caused by intramolecular hyperfine interaction (interaction between electron spin and nuclear spin). Hyperfine interactions are classified into Fermi contact interactions involving s electrons and magnetic dipole interactions involving p, d, and f electrons. Even in the case of a single molecule isolated in vacuum, very small hyperfine interactions can occur if the relativistic and non-adiabatic effects are considered. Nonetheless, the ortho-para conversion rate due to this mechanism is predicted to have the timescale comparable to that of the lifetime of the universe, and thus can be ignored [12]. Since this literature emphasized the interaction of hydrogen with surfaces, the authors suggest to the interested readers to refer to other reference materials specifically devoted to ortho-para conversion such as Reference [10] and other references therein. During ortho-para conversion, the approaching hydrogen molecule transfers excess rotational energy to

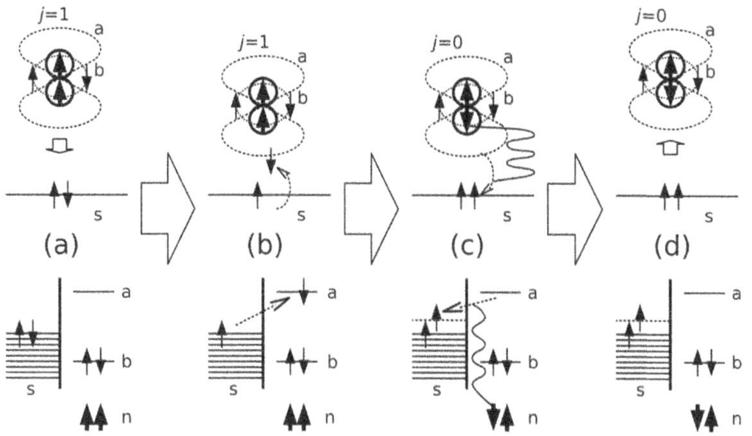

Fig. 3.13 Schematic diagram of ortho-para conversion caused by hyperfine interaction on the surface (The UY process). The upper diagram shows the space diagram and the lower diagram shows the energy diagram. **a** shows the incident ortho hydrogen, with rotational state $j = 1$, **b** shows the virtual electron transition from the surface to the molecule, **c** shows the movement of electron from the molecule to the surface with electron-nuclear spin inversion due to hyperfine interaction. **d** shows the desorption of parahydrogen, with rotational state $j = 0$. Thin arrows represent electron spins, and thick arrows represent nuclear spins. *s* is the electronic state of the surface, *b* is the bonding level (orbital) of the molecule, *a* is the antibonding level (orbital), and *n* is the nuclear spin Reprinted with permission from [55]. Copyright 2013 by Asakura Publishing

the surface via Coulomb interaction (between electrons and electrons, and between electrons and nuclei) [13].

The mechanism of ortho-para conversion on the surface caused by hyperfine interaction is shown in Fig. 3.13. When ortho hydrogen at the electronic ground state (its two electrons are in the bonding orbital) approaches the surface, an overlap occurs between (process (a)) the antibonding orbital of the molecule and the surface electron orbital. This causes an electron transition from the surface to the molecule (process (b)). Since this state is a metastable state that deviates from the equilibrium condition, it quickly shifts to a stable state. If the extra electrons acquired by H_2 induce hyperfine interactions due to field distortion, a reversal of electron spin and nuclear spin accompanied by rotational cooling occurs to preserve the spin of the entire system (process (c)). At the same time, the surface receives the energy released by rotational cooling as the electrons move from the antibonding orbital to the surface orbital. As a result, H_2 is converted to parahydrogen and desorbs from the surface (process (d)). Processes (b) and (c) are rare events, so the conversion rate is in the order of seconds even for ferrimagnetic and antiferromagnetic surfaces.

As discussed in the previous section, the hindered rotation of the H_2 has a preferential orientation with respect to the surface, in contrast to the isotropic rotation in the gas phase. It turns out that the orientation of the hydrogen molecule axis affects the conversion rate. When the molecular axis is perpendicular to the surface, the difference in the field perceived by the two nuclei of the molecule increases and the hyperfine interaction increases, thus increasing the conversion speed. On the other hand, when the molecular axes are parallel to the surface, the hyperfine interaction is small, and the conversion rate is low. The Fermi contact interaction $H_{\mathrm{HC}} \propto \vec{I} \cdot \vec{S}\delta\left(\vec{r} - \vec{R}\right)$ mainly contributes to the hyperfine interaction with the starting condition that the molecule and the surface are in the electronic ground state. Here, $\vec{r}$ and $\vec{S}$ are the electron coordinate and spin operators, $\vec{R}$ and $\vec{I}$ are the nuclear coordinate and spin operators, respectively, and δ is the Dirac delta function. Process (b) in Fig. 3.13 is a transition, due to Coulomb interaction H_{C} that acts on electrons, from their initial state Ψ_i with energy E_i to the intermediate state Ψ_I with energy E_I. Process (c) describes a transition from Ψ_I to final state Ψ_f with energy E_f, due to Fermi contact interaction H_{HC}, where the ortho-para conversion yield $W_{\mathrm{o}\to\mathrm{p}}$ is given by

$$W_{\mathrm{o}\to\mathrm{p}} = \frac{2\pi}{\hbar} \sum_f \left| \sum_I \frac{\langle \Psi_f | H_{\mathrm{HC}} | \Psi_I \rangle \langle \Psi_I | H_{\mathrm{C}} | \Psi_i \rangle}{E_i - E_I} \right|^2 \delta\left(E_i - E_f\right). \tag{3.34}$$

δ is the Dirac delta function.

As an example, consider the case where a $j = 1$ ortho hydrogen approaches the B-terminated (001) surface of ABO_3 perovskite. Suppose the surface B atoms have d orbitals extending in the direction perpendicular to the surface, which interact with hydrogen molecules. The averages of the Fermi contact interaction and Coulomb interaction with respect to the electronic coordinates, spin, and nuclear spin are expressed as $\langle H_{\mathrm{HC}}(Z,\theta)\rangle$ and $\langle H_{\mathrm{C}}(Z,\theta)\rangle$, respectively. These expressions are functions of the center-of-mass of the molecule, the distance of the molecule from the surface Z, and the inclination angle θ of the molecular axis with respect to the surface normal. Coulomb interaction $\langle H_{\mathrm{C}}(Z,\theta)\rangle$ can be further divided into attractive electron-nucleus interaction component $\langle H_{\mathrm{CA}}(Z,\theta)\rangle$ and repulsive electron-electron interaction component $\langle H_{\mathrm{CR}}(Z,\theta)\rangle$.

Figure 3.14a shows the average Fermi contact interaction for different molecular axis orientation and molecule distance from the surface, $\langle H_{\mathrm{HC}}(Z,\theta)\rangle$. As θ approaches 0°, the hydrogen molecule axis becomes perpendicular to the surface, and the magnetic field difference between the two nuclei increases. With this, the Fermi contact interaction also increases. Conversely, when the axis approaches 90° such that the molecule is oriented parallel to the surface and the magnetic field between the two nuclei becomes symmetric, the Fermi contact interaction becomes smaller. Figure 3.14b shows the mean $\langle H_{\mathrm{CA}}(Z,\theta)\rangle$ of the attractive Coulomb interaction. When the hydrogen molecule approaches the surface from a distance $Z \to \infty$, the energy of the interaction decreases first because the hydrogen nuclei are attracted by the surface d-electrons. As the molecule moves closer to the surface, there will be a minimum in the potential energy profile. In this case, the minimum point $Z^{\mathrm{A}}_{\mathrm{min}}(\theta)$ corresponds to the configuration in which the hydrogen nuclei are oriented such that the overlap of their orbitals with the surface d electron density is maximized. For

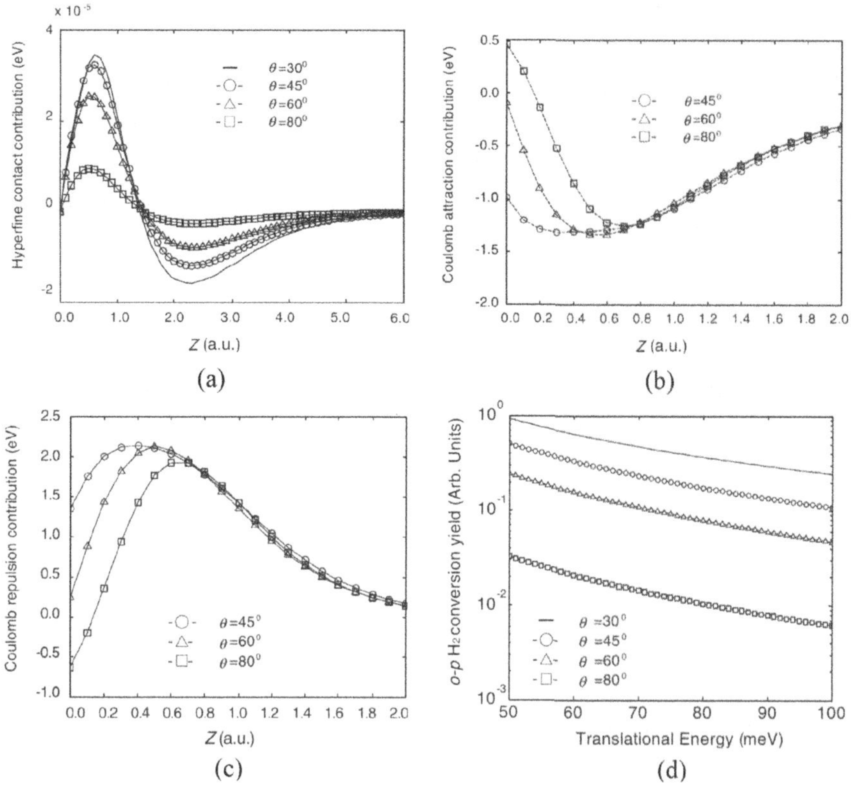

Fig. 3.14 $j = 1$ of ortho hydrogen near the B-terminated surface of ABO_3 perovskite (001) : contributions of **a** Fermi contact interaction $\langle H_{\rm HC}(Z,\theta)\rangle$, **b** attractive Coulomb interaction $\langle H_{\rm CA}(Z,\theta)\rangle$ acting on electron as, and **c** Coulomb repulsion interaction $\langle H_{\rm CR}(Z,\theta)\rangle$ as functions of the molecule distance from the surface (Z) and the molecular axis relative to the surface normal (θ), and **d** dependence of ortho-para conversion yield $W_{\rm o\to p}$ on incident translational energy. Reprinted with permission from [14]. Copyright 2001 by the Physical Society of Japan

$Z < Z^{\rm A}_{\rm min}(\theta)$, the contribution of the antiphase part of the surface d-electron orbital appears as a repulsive component, especially when the inclination of the molecular axis is close to the surface plane, and cancels out the attractive component at a sufficiently close distance. The region $Z > Z^{\rm A}_{\rm min}(\theta)$ takes slightly lower energy due to the weak hybridization between the molecular antibonding orbital and the surface electron orbitals. This is more obvious when the molecular axis is oriented almost parallel to the surface normal as compared when it is almost parallel to the surface plane. However, this is related to the tendency of diatomic molecules to be adsorbed perpendicular to the surface. Figure 3.14c shows the mean $\langle H_{\rm CR}(Z,\theta)\rangle$ of the repulsive Coulomb interaction. This Z dependence and its mechanism correspond to the inverted version of Fig. 3.13 c. The probability of electron transfer from the surface to the molecule in process (b) is higher as the absolute value of the attractive and repulsive components of Coulomb interaction is higher. The potential energy governing the molecular motion corresponds to the sum of the attractive and repulsive components of the Coulomb interaction, so the effective interaction region is close to the physical adsorption distance $\left[Z \gg Z^{\mathrm{A}}_{\mathrm{min}}(\theta)\right]$. The ortho-para conversion yield $W_{\mathrm{o}\to\mathrm{p}}$ given by (3.41) is shown in Fig. 3.14d as a function of the incident translational energy. The conversion yield decreases exponentially with increasing incident translational energy, which reflects a decrease in contact time with the surface due to increased velocity. The conversion yield tends to increase when the tilt of the molecular axis is close to the surface normal, which is due to the Fermi contact interaction and to the remarkable dependence with θ.

3.3 Quantum Molecular Dynamics of Hydrogen on Solid Surfaces: Formalism

The previous section discussed various phenomena related to hydrogen atoms and molecules adsorb on surfaces. Whether the adsorption (and other related phenomena such as reflection, desorption, etc.) could occur depends on various factors, notably on the PES of the hydrogen molecule on the surface, as well as on the translational, rotational, and vibrational states of the impinging hydrogen molecules, and on the temperature of the surface. The atomic scale study of these phenomena is directly related to quantum molecular dynamics, which is the focus of discussion in this section. Due to the quantum mechanical nature of the hydrogen nuclei, the problem of predicting the outcome from the quantum molecular dynamics can be transformed into the scattering problem in quantum mechanics. In this section, necessary tools will be developed, and phenomena related to quantum dynamics of hydrogen atoms and molecules will be discussed.

To begin the discussion, consider the Hamiltonian of H_2-surface system, in which, the internal degrees of freedom of the molecule are written in spherical coordinate, and the other degrees of freedom are expressed in Cartesian coordinate:

$$\begin{aligned} H(X, Y, z; r, \theta, \phi) = &-\frac{\hbar^2}{2M_{mol}}\left(\frac{\partial^2}{\partial X^2}+\frac{\partial^2}{\partial Y^2}+\frac{\partial^2}{\partial Z^2}\right)\\ &-\frac{\hbar^2}{2\mu}\left(\frac{1}{r}\frac{\partial^2}{\partial r^2}r\right)+\frac{1}{2\mu r^2}\mathbf{L}^2\\ &+U(X, Y, Z, r, \theta, \phi). \end{aligned} \tag{3.35}$$

The above Hamiltonian describes the same system, similar with that of (3.14), except that different coordinate systems were utilized. For H_2 adsorption on surfaces, it is important to consider the distance of the CM of H_2 from the surface (Z) and the H-H interatomic distance (r). In practice, the other degrees of freedom can usually be treated as parameters. For example, both X and Y can be regarded as the adsorption sites on the surface, while both θ and ϕ are related to the rotational states of the H_2, similar to the case of hindered rotation discussed earlier.

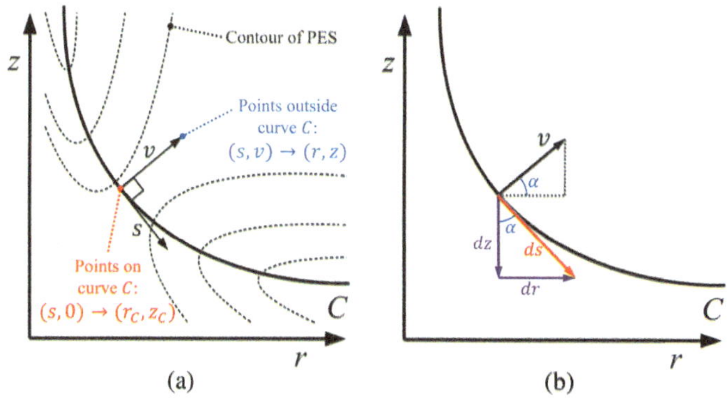

Fig. 3.15 Schematic diagrams showing coordinates transformation from the molecule-surface coordinate (r, z) to the reaction path coordinate (s, v) on the two-dimential PES, for constant values of (X, Y, θ, ϕ). Here, the effective one-particle coordinate z is defined as $z = Z\sqrt{M/\mu}$. Figure 3.15a shows the curve C passing through the minimum energy path of the PES, where s is the coordinate parallel to the C, and v is the coordinate perpendicular to C. The vectors s and v forms Euclidean space *locally* along C. Figure 3.15b shows the relation of the differential quantities ds, dr, and dz on the same curve C, as well as the definition of the angle α. Note that ds and v are orthogonal. Adapted with permission from [55]. Copyright 2013 by Asakura Publishing

The next step is to transform (r, Z) to the reaction coordinate (s, v), as shown in Fig. 3.15a. To do so, the coordinate Z is first expressed in the effective one-particle coordinate by using the transformation $z = Z\sqrt{M_{mol}/\mu}$. The Hamiltonian written in this new coordinate z then takes the form:

$$\begin{aligned} H(X, Y, z; r, \theta, \phi) = &-\frac{\hbar^2}{2M_{mol}}\left(\frac{\partial^2}{\partial X^2} + \frac{\partial^2}{\partial Y^2}\right) \\ &-\frac{\hbar^2}{2\mu}\left(\frac{\partial^2}{\partial z^2} + \frac{1}{r}\frac{\partial^2}{\partial r^2}r\right) + \frac{1}{2\mu r^2}\mathbf{L}^2 \\ &+ U(X, Y, z, r, \theta, \phi). \end{aligned} \tag{3.36}$$

To transform the coordinate (r, z) to (s, v), consider their relation as shown in Fig. 3.15a. With reference to the vector v in the figure, we can obtain the following:

$$r = r_C(s) + v \cdot \cos(\alpha) \tag{3.37}$$

$$z = z_C(s) + v \cdot \sin(\alpha). \tag{3.38}$$

Next, consider the differential quantities ds, dr, and dz in Fig. 3.15b to construct two additional equations:

$$\sin(\alpha) = \frac{\partial r_C(s)}{\partial s} \tag{3.39}$$

$$\cos(\alpha) = -\frac{\partial z_C(s)}{\partial s}. \tag{3.40}$$

Then, relate the above four equations using the curvature defined by:

$$\kappa(s) \equiv \frac{\partial \alpha}{\partial s} \tag{3.41}$$

From the definition of the curvature in (3.41), substitute the expressions for $\sin(\alpha)$ from (3.39) and $\cos(\alpha)$ from (3.40). The curvature then takes the form:

$$\kappa(s) = \underbrace{\frac{\partial^2 r_c(s)}{\partial s^2}\left[\sqrt{1-\left(\frac{\partial r_c(s)}{\partial s}\right)^2}\right]^{-1}}_{\text{from eq.(3.39)}} = \underbrace{\frac{\partial^2 z_c(s)}{\partial s^2}\left[\sqrt{1-\left(\frac{\partial z_c(s)}{\partial s}\right)^2}\right]^{-1}}_{\text{from eq.(3.40)}}. \tag{3.42}$$

Lastly, calculate the Jacobian determinant $\eta(s, v)$:

$$\eta(s, v) = \left|\frac{\partial(r, z)}{\partial(s, v)}\right| = \begin{vmatrix} \frac{\partial r}{\partial s} & \frac{\partial r}{\partial v} \\ \frac{\partial z}{\partial s} & \frac{\partial z}{\partial v} \end{vmatrix} = 1 - v\kappa(s). \tag{3.43}$$

With these, the Hamiltonian given in (3.36) can be written as:

$$\begin{aligned} H(X, Y, z; r, \theta, \phi) = &-\frac{\hbar^2}{2M_{mol}}\left(\frac{\partial^2}{\partial X^2} + \frac{\partial^2}{\partial Y^2}\right) \\ &- \frac{\hbar^2}{2\mu}\left(\frac{1}{\eta(s, v)}\frac{\partial}{\partial s}\frac{1}{\eta(s, v)}\frac{\partial}{\partial s} + \frac{1}{\eta(s, v)}\frac{\partial}{\partial v}\eta(s, v)\frac{\partial}{\partial v}\right) \\ &+ \frac{1}{2\mu r^2}\mathbf{L}^2 + U(X, Y, z, r, \theta, \phi). \end{aligned} \tag{3.44}$$

The next step is to solve the Hamiltonian. The central idea is to treat the scattering of the hydrogen wavefunction with the potential $U(X, Y, z, r, \theta, \phi)$. To solve such scattering problem, one usually transforms the Schrödinger equation in reaction coordinate (3.44) to new equation(s) such that it is easier to solve the problem. Nonetheless, regardless of the methods used, the results from different methods should explain the same physics, and the choice of methods is often a matter of training and personal taste [15].

From here, the time-dependent perturbation theory will be used to solve (3.44). The idea is to consider the Hamiltonian *without* the barrier as the unperturbed Hamiltonian H_0, and treat the barrier as the perturbation. Consider the dynamics of dissociative adsorption of molecules on the surface. The time-dependent Schrödinger equation is

$$\left[\mathrm{i}\hbar\frac{\partial}{\partial t} - H(s, v, X, Y, \theta, \phi)\right]\Psi(s, v, X, Y, \theta, \phi, t) = 0. \tag{3.45}$$

From the spectral theorem, the solution Ψ can be written as the linear combination of eigenfunction $\Psi_m(s, v, X, Y, \theta, \phi)$:

$$\Psi(s, v, X, Y, \theta, \phi, t) = \sum_m C_m \mathrm{e}^{-\mathrm{i}E_m t/\hbar}\Psi_m(s, v, X, Y, \theta, \phi). \tag{3.46}$$

Here, the index m represents the quantum number, and E_m represents the eigenenergy for the quantum number m. C_m is an arbitrary coefficient. Since $H(s, v, X, Y, \theta, \phi)$ does not depend on time, $\Psi(s, v, X, Y, \theta, \phi, t)$ represents a stationary wave packet. Now, assume that the potential energy on the reaction path takes a constant value in $s \to \pm\infty$, as shown in Fig. 3.16. Since the molecules are isolated in vacuum as $s \to -\infty$ (defined as V), the interaction between molecule and surface vanishes. Thus, the internal state $(s, v, X, Y, \theta, \phi)$ no longer depends on s. Likewise, as $s \to +\infty$ (defined as S), the molecules dissociate and move freely relative to one another on the surface, so the internal state should also be largely independent of s. Therefore, the internal state depends on s only in the region where state transition occurs due to the change in potential energy. The wave function Ψ^V in the region V is

$$\left[i\hbar\frac{\partial}{\partial t} - H^V(v, X, Y, \theta, \phi)\right]\Psi^V(s, v, X, Y, \theta, \phi, t) = 0 \tag{3.47}$$

for the Hamiltonian H_a^V for free molecules. Similarly, it is assumed that the wave function Ψ^S in the region S is also given as a solution to the Schrödinger equation

$$\left[i\hbar\frac{\partial}{\partial t} - H^S(v, X, Y, \theta, \phi)\right]\Psi^S(s, v, X, Y, \theta, \phi, t) = 0 \tag{3.48}$$

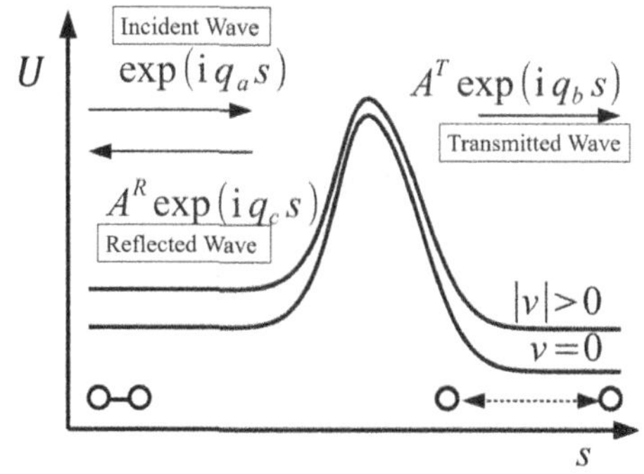

Fig. 3.16 Incident, reflected, and transmitted waves along the reaction path C, as defined in Fig. 3.15. The figure shows the incident [$\exp(iq_a s)$], reflected [$\exp(iq_c s)$], and transmitted [$\exp(iq_b s)$] waves. Adapted with permission from [55]. Copyright 2013 by Asakura Publishing

for the Hamiltonian H^S for two free atoms in surface two-dimensional space. Since H^V and H^S do not depend on s, the s dependence of Ψ^V and Ψ^S is a plane wave and can be written as

$$\Psi^V(s, v, X, Y, \theta, \phi, t) = \sum_q \sum_m C^V_{q,m} e^{-i(E^V_q + E^V_m)t/\hbar} e^{iqs} \Phi^V_m(v, X, Y, \theta, \phi) \quad (3.49)$$

$$\Psi^S(s, v, X, Y, \theta, \phi, t) = \sum_q \sum_m C^S_{q,m} e^{-i(E^S_q + E^S_m)t/\hbar} e^{iqs} \Phi^V_m(v, X, Y, \theta, \phi) \quad (3.50)$$

However, Φ^V and Φ^S are generalized internal wave functions that depend on spatial degrees of freedom other than s. In addition, the terms q, E^V_q and E^S_q in the sum are generalized translational wave numbers and their corresponding energies in the limit regions V and S. On the other hand, m, E^V_m and E^S_m are generalized internal quantum numbers and energies in the limit regions V and S. $C^V_{q,m}$ and $C^S_{q,m}$ are coefficients that depend on the combinations of q and m.

The transition from the molecular to dissociated state occurs at the bending of curve C in Fig. 3.15, which results in a scattering problem in the reaction path coordinate system. In the scattering region where the s dependence of the potential energy is significant, the generalized translational motion is deviated from the plane wave, and the internal wavefunction depends on s. Therefore, following (3.13) and (3.14), equation (3.10) can be rewritten as

$$\Psi(s, v, X, Y, \theta, \phi, t) = \sum_q \sum_m C_{q,m} e^{-iE_{q,m}t/\hbar} \psi_{q,m} \Phi_{q,m}(s; v, X, Y, \theta, \phi) \quad (3.51)$$

using the generalized translational wave function $\psi_m(s)$ and the internal wave function $\Phi_m(s; v, X, Y, \theta, \phi)$. Here, $\psi_{q,m}\Phi_{q,m}(s; v, X, Y, \theta, \phi)$ is an eigenfunction corresponding to the quantum number (q, m), and its eigenvalue is $E_{q,m}$. In the limit of $s \to \pm\infty$, the eigenfunction is considered to have an asymptotic form of

$$\psi_{q,m}(s) \to \psi_q(s) = e^{iqs} (s \to \pm\infty) \quad (3.52)$$

$$\Phi_m(s; v, X, Y, \theta, \phi) \to \Phi_m(v, X, Y, \theta, \phi)(s \to \pm\infty) \quad (3.53)$$

Here, the probability amplitude $A^T_{b,a}$ in which the incident wave Ψ^V_a with wavenumber q_a and internal quantum number m_a is transmitted through the scattering region, and yields the generalized translational wavenumber q_b and wavefunction Ψ^S_b with internal quantum number m_b after the scattering region (see Fig. 3.16). The wavefunctions of the incident wave and the transmitted wave are denoted respectively by

$$\Psi^V_a(s, v, X, Y, \theta, \phi, t) = e^{-i(E^V_{qa} + E^V_{ma})t/\hbar} \psi_{q_a}(s) \Phi_{m_a}(v, X, Y, \theta, \phi) \quad (3.54)$$

$$\Psi_b^S(s, v, X, Y, \theta, \phi, t) = \mathrm{e}^{-\mathrm{i}\left(E_{q_b}^S + E_{m_b}^S\right)t/\hbar}\psi_{q_b}(s)\Phi_{m_b}(v, X, Y, \theta, \phi) \tag{3.55}$$

When the incident wave reaches the scattering region, it experiences the potential change $W^V = H_a - H_a^V$. Here, it is assumed that this potential change can be treated as a perturbation. That is, the wave function Ψ_a in the scattering region can be described by a linear combination in the functional system in the region V. For the sake of brevity, the wavefunction $\Psi_a^V(s, v, X, Y, \theta, \phi, t)$ is rewritten in the abstract vector space as

$$\left|\Psi_a^V(t)\right\rangle = \mathrm{e}^{-\mathrm{i}(E_{qa}^V + E_{ma}^V)t/\hbar}\left|\Psi_{q_a,m_a}^V\right\rangle \tag{3.56}$$

From the first-order time-dependent perturbation theory, the wave function for scattering by a perturbation W^V (first-order perturbation) is represented by

$$\begin{aligned}\left|\Psi_a^V(t)\right\rangle^{(1)} &= \frac{1}{\mathrm{i}\hbar}\sum_{q',m'} \mathrm{e}^{-\mathrm{i}\left(E_{q'}^V + E_{m'}^V\right)t/\hbar}\left|\Psi_{q',m'}^V\right\rangle \\ &\times \int_{t_0}^{t} dt' \mathrm{e}^{\mathrm{i}\left[\left(E_{q'}^V + E_{m'}^V\right) - \left(E_{qa}^V + E_{ma}^V\right)\right]t'/\hbar}\left\langle\Psi_{q',m'}^V\right|W^V\left|\Psi_{q_a,m_a}^V\right\rangle\end{aligned} \tag{3.57}$$

Here, t_0 indicates the time when scattering starts to occur, and $\left\langle\Psi_{q',m'}^V\right|W^V\left|\Psi_{q_a,m_a}^V\right\rangle$ indicates the transition matrix element due to the perturbation W^V. The above equation expresses the wave function after scattering as a wave packet of the function system $\left|\Psi_{q,m}^V\right\rangle$ in the region V. Similarly, when scattering occurs twice (secondary perturbation), the wave function is represented as

$$\begin{aligned}\left|\Psi_a^V(t)\right\rangle^{(2)} &= \left(\frac{1}{\mathrm{i}\hbar}\right)^2 \sum_{q'',m''} \mathrm{e}^{-\mathrm{i}\left(E_{q''}^V + E_{m''}^V\right)t/\hbar}\left|\Psi_{q'',m''}^V\right\rangle \\ &\times \sum_{q',m'} \int_{t_0}^{t} dt'' \mathrm{e}^{\mathrm{i}\left[\left(E_{q''}^V + E_{m''}^V\right) - \left(E_{q'}^V + E_{m'}^V\right)\right]t''/\hbar}\left\langle\Psi_{q'',m''}^V\right|W^V\left|\Psi_{q',m'}^V\right\rangle \\ &\times \int_{t_0}^{t} dt' \mathrm{e}^{\mathrm{i}\left[\left(E_{q'}^V + E_{m'}^V\right) - \left(E_{qa}^V + E_{ma}^V\right)\right]t'/\hbar}\left\langle\Psi_{q',m'}^V\right|W^V\left|\Psi_{q_a,m_a}^V\right\rangle\end{aligned} \tag{3.58}$$

If the potentials of the regions V and S are the same, the eigenfunction systems in both regions are similar. At this time, based on Lippmann-Schwinger equation [16, 17], summing up to the infinite perturbation term, the probability that the wave transmitted through the scattering region is observed as Ψ_b^S with amplitude $A_{b,a}^T$ that can be derived as

$$
\begin{aligned}
A^T_{b,a} &= \langle \Psi_b(t)|\Psi_a(t)\rangle = \left\langle \Psi^S_{q_b,m_b}|\Psi^V_{q_a,m_a}\right\rangle \\
&+ \frac{1}{\mathrm{i}\hbar} \lim_{t_0\to-\infty} \lim_{t_1\to+\infty} \int_{t_0}^{t_1} dt' \mathrm{e}^{\mathrm{i}\left[\left(E^S_{q_b}+E^S_{m_b}\right)-\left(E^V_{q_a}+E^V_{m_a}\right)\right]t'/\hbar} \left\langle \Psi^S_{q_b,m_b}|W^V|\Psi_a(0)\right\rangle \\
&= \delta_{b,a} - 2\pi\mathrm{i}\delta\left(\left(E^S_{q_b}+E^S_{m_b}\right)-\left(E^V_{q_a}+E^V_{m_a}\right)\right)T^T_{b,a}
\end{aligned}
\tag{3.59}
$$

Here, $T^T_{b,a} = \left\langle \Psi^S_{q_b,m_b}\middle|W^V|\Psi_a(0)\right\rangle$ is a transition matrix element from the incident wave Ψ^V_a to the transmitted wave Ψ^S_b. Since the Hamiltonian H does not depend on time, the time dependence of $\Psi_a(t)$ is canceled on the bra and the ket sides, resulting in a time-independent expression. The first term $\delta_{b,a}$ is Kronecker delta, which represents the elastic scattering. The second term, $\delta(E)$, is the Dirac delta function, which represents the requirement for energy conservation before and after scattering. That is, if the internal energy changes from $E^V_{m_a}$ to $E^S_{m_b}$, the translational kinetic energy changes from $E^V_{q_a}$ to $E^S_{q_b}$ so as to maintain the energy conservation. Likewise, the probability amplitude $A^R_{b,a}$ and the transition matrix element $T^R_{b,a}$ observed can also be derived as a reflected wave.

In the dissociative adsorption process, the contribution of the elastic scattering term usually does not appear because there is a difference in the potential depth and the potential shape in the v direction between regions V and S. Also, the potential energy of region S should be influenced by the periodic structure of the surface considered. Depending on the shape of the potential curve, bound states may appear, and the theory presented above may break down. That is, there is no guarantee that the eigenfunction system in regions V and S has completeness even in the scattering region. These difficulties can be avoided by performing numerical calculations based on the coupled channel method [18–21]. As the coupled channel method directly solves (3.44), the assumption based on the perturbation theory is not necessary.

3.4 Quantum Molecular Dynamics of Hydrogen on Solid Surfaces: Phenomena

In this section, phenomena related to the quantum molecular dynamics on solid surfaces will be discussed, in which the results are obtained based on the generalized scattering model in reaction coordinate presented in the last section. Figure 3.17 shows the schematic diagram of some of possible reactions that could occur from the scattering of the diatomic molecule from the solid surface where some of the reactions are briefly discussed in the first chapter. The discussion of the phenomena will be based on the point-of-view of generalized reaction coordinate. The use of generalized reaction coordinate will allow the understanding of the role of the energy transfer among translation, rotation, and vibration energies, and consequently their roles in assisting/inhibiting certain outcomes such as adsorption, desorption, etc. In the following subsections, the discussion was arranged so that the phenomena that

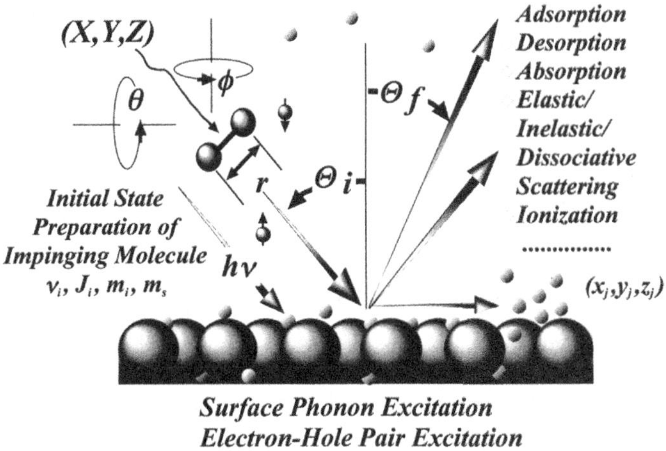

Fig. 3.17 Schematic diagram of some of possible reactions that could occur from the impinging diatomic molecule. The coordinates $(X, Y, Z, r, \theta, \phi)$ characterize the positions of the diatomic molecule (see definition from Fig. 3.7).The initial states of the impinging molecule can be characterized by the dynamical variables (v_i, J_i, m_i, m_s), which denote the initial vibrational, rotational, azimuthal, and spin quantum states, respectively. The angles Θ_i and Θ_f correspond to the incident and scattering angles of the diatomic molecule with respect to the surface normal, respectively. Lastly, the locations of the constituent atoms of the diatomic molecule after scattering are indicated by the coordinate $\left(x_j, y_j, z_j\right)$. Reprinted with permission from [22]. Copyright 2000 by Elsevier

can be adequately explained by simplified model are presented first, followed by other phenomena that require more generalized models.

3.4.1 Tunneling Effect

For the tunneling effect in dissociative adsorption, one-dimensional coordinate s (as defined in Fig. 3.15) is qualitatively adequate in many cases. As an example, consider the dissociative adsorption of an impinging hydrogen molecule on the surface. To proceed from molecular state in the gas phase to the dissociative adsorption on the surface, the hydrogen molecule must overcome the activation barrier represented by the potential $V(s)$ along the reaction path s.

When nuclear motion is treated classically, and that the incident energy E_t is lower than the potential barrier E_a, the incident molecule is always reflected, and adsorption does not occur. Conversely, if incident energy is higher than the barrier, it always crosses the barrier and is adsorbed. In other words, E_a gives an exact threshold, and the transmission probability T becomes 0 when $E_t < E_a$ and 1 when $E_t > E_a$. On the other hand, when nuclear motion is treated quantum mechanically, the incident wave passes through the barrier with a finite probability. Figure 3.18a schematically shows two cases of the incident wavefunctions, one with higher and

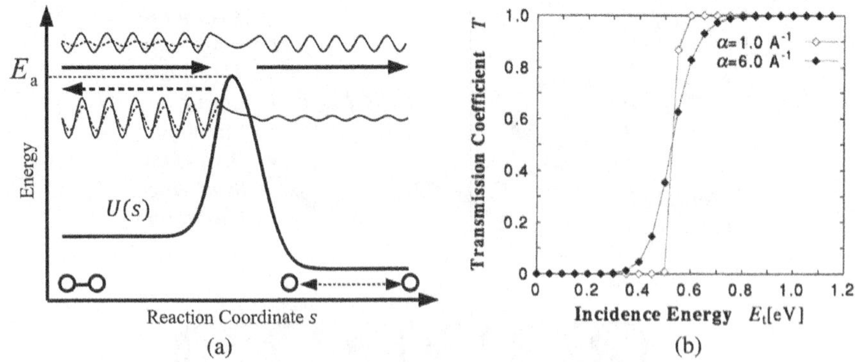

Fig. 3.18 Tunneling effect in dissociative adsorption. **a** Conceptual diagram of reaction path coordinate s. The thick curve represents the potential energy curve $V(s)$ with the barrier E_a. The solid line represents the incident wave and the transmitted wave, while the dashed line represents the reflected wave. The upper wave shows the case of the incident wave whose energy is higher than the potential barrier, and the lower wave shows the case of the incident wave of lower energy. **b** Dependence of the potential barrier transmittance T on the incident kinetic energy E_t and the barrier thickness controlled by parameter α [22, 23]. **a** was adapted with permission from [55]. Copyright 2013 by Asakura Publishing. **b** was reprinted from [23]. Copyright 1995 by the Physical Society of Japan

one with lower energies with respect to the potential barrier. In the case of $E_t < E_a$, the wavefunction inside the barrier region has the form of an exponentially decaying wavefunction. As a result, the incident wavefunction is mostly reflected, yet there are non-zero wave functions transmitted through the barrier. This is the tunneling effect in dissociative adsorption. Conversely, if $E_t > E_a$, the wave function in the barrier region is elongated and locally scattered, in which some components of the wavefunction are reflected. In this case, the transmission probability decreases as compared to the classical case. Figure 3.18b shows the transmission probability T as a function of the incident energy E_t based on the coupled channel method, in which the potential barrier thickness is controlled by the parameter α (large values of $|\alpha|$ correspond to thin barriers, and vice versa). Here, the potential of the activation barrier is given by $V(s) = E_a/\cosh^2(\alpha s)$, and E_a is at 0.54 eV, which resembles that of the copper surface. At the classical mechanical threshold E_a, incidence and reflection occur with equal probability, and the transmission probability T is 0.5. Before and after E_a, the transmission probabilities converge smoothly toward classical dynamics (0 and 1). When the barrier is thick ($\alpha = 0.1$ Å^{-1}), the behavior is similar to that of classical mechanics, but when it is thin ($\alpha = 0.6$ Å^{-1}), tunneling and anti-tunneling appear prominently.

3.4.2 *Effects of Vibration and Rotation on Dissociative Adsorption*

In addition to the reaction along the reaction coordinate s, the effect due to the internal states (ν, θ, ϕ) should be considered. The effects of vibration (represented by ν in the PES from Fig. 3.15) and rotation (represented by θ and ϕ) states of molecules in dissociative adsorption and association desorption described in Chap. 1 can be quantitatively shown by quantum dynamics calculation based on the coupled channel method. As an example, Fig. 3.19 shows the results of the dissociative adsorption probability of deuterium (D_2) on Cu(111). This figure shows the dependence on the translational energy E_t, the vibrational quantum number ν and the rotational quantum number J of the incident molecule. However, the sum is taken for the magnetic quantum number m (representing the direction of the rotation axis).

First, consider the adsorption probability as a function of the rotational quantum number J. The most notable behavior from the Fig. 3.19 is that the adsorption probability is not always increasing with respect to the higher values of J. This behavior is different from the case of the adsorption probability as the function of E_t discussed previously in Fig. 3.21, which shows that the adsorption probability always increases with higher values of E_t. From Fig. 3.19, except for the case of the relatively low translational energy E_t, the adsorption probability initially decreases as J increases from to 5 or 6. As J increases further from 6 or 7 to 14, the adsorption probability increases due to the rotational-translational energy transfer effect. At relatively low values of J (roughly $J < 6$), the so-called steering effect dominates [23]. Briefly, the steering effect is a phenomena where the approaching molecule reorients itself upon approaching the surface. In this case, there are two possible factors that contribute to the lowering of adsorption probability. The first one is due to the relatively short time that the molecule spends in a favorable orientation. The second one is when the molecule first approaches the surface with unfavorable orientation, it uses up its translational energy to reorient itself. On the other hand, at relatively high values of J (roughly $J > 6$), the molecule has enough rotational energy, and the rotational-translational energy transfer effect dominates [23]. Back

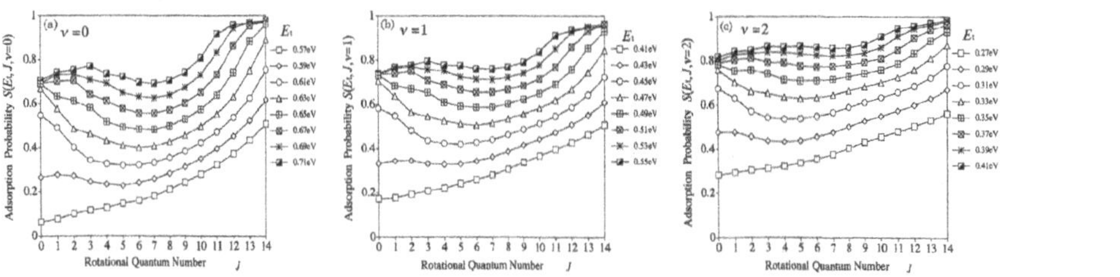

Fig. 3.19 Dependence of adsorption probability on translational energy E_t, vibrational quantum number ν and rotational quantum number J in dissociative adsorption of deuterium on Cu(111) (reprinted from [24]) in the case of **a** $\nu = 0$, **b** $\nu = 1$, and **c** $\nu = 2$, respectively. Reprinted with permission from [24]. Copyright 1999 by the Physical Society of Japan

in Chap. 1, the three-dimensional PES as the function of the reaction path and the angle of the approaching molecule is shown in Fig. 1.9. With high value of J, the approaching molecule has enough rotational energy for a wide range of $\cos(\theta)$, and the approaching molecule does not require to convert translational energy along the reaction path s to the rotational energy along the $\cos(\theta)$ axis to overcome the barrier. Here, the rotational-translational energy transfer occurs because along the different reaction path coordinates s, the optimal hydrogen molecule bond lengths r are different. Since rotational constant depends on the bond length r, there is a coupling between rotational and translation motions where the rotational energy is converted to translational energy via bond length extension process (see Fig. 3.18 of [22]).

Next, consider the ν dependency. In general, higher values of ν increases the adsorption probability for the same values of (E_t, J) in the so-called vibrational-assisted adsorption effect [25]. The vibrational-assisted adsorption effect tends to be especially remarkable in the region where E_t and J are small. On the other hand, if both E_t and J are large, sufficient adsorption probability can be obtained only by the original translational motion and the energy transfer between rotation and translation, so that the assistance by vibration becomes unnecessary and the ν dependence becomes smaller. Also, when $\nu = 1$ and 2 are compared with $\nu = 0$, it is found that the increase in ν tends to reduce the J dependence. This indicates that for large ν, the vibration-assisted adsorption effect dominates the steering effect and the energy transfer between translation and rotation.

Next, examine the m dependence of the adsorption probability at $J = 6$ at which the steering effect disappears (see Fig. 3.20). For the cartwheel-like rotation, which corresponds to $m = 0$, the rotation axis is parallel to the surface (thus, there is no surface perpendicular component of the magnetic moment created by atomic motion). On the other hand, for the helicopter-like rotation, which corresponds to $|m| = J = 6$, the surface perpendicular component of the magnetic moment is at maximum (for the current value of $J = 6$; note that $|m|$ must be equal or lower than J). The overall tendency is that the larger the value of $|m|$ (the closer to a helicopter-like), the higher the probability of adsorption, and the potential barrier for adsorption is shifted down

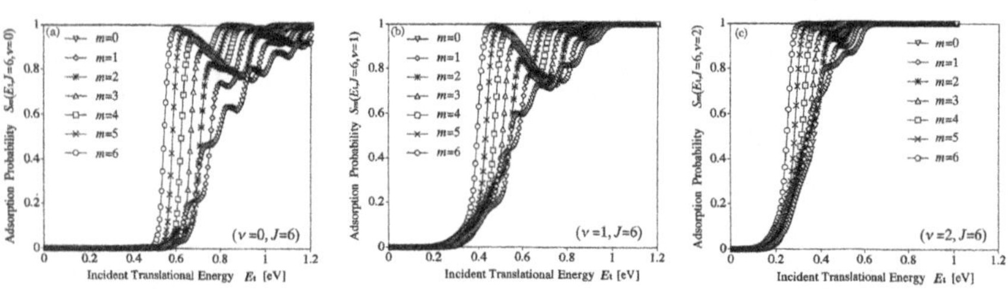

Fig. 3.20 Dependence of the adsorption probability on the dissociative adsorption of deuterium on Cu(111) on translational energy E_t, vibrational quantum number ν and magnetic quantum number m [24]. When **a** $\nu = 0$, (b) $\nu = 1$, and **b** $\nu = 2$, respectively. Reprinted with permission from [24]. Copyright 1999 by the Physical Society of Japan

in energy, allowing the approaching molecule with lower value of E_t to adsorb. This is because when $|m|$ is small (close to cartwheel-like), atomic arrangement in which the molecular axis becomes perpendicular to the surface occurs frequently. The PES from Fig. 1.9 shows that the perpendicular orientation is a relatively unfavorable orientation, and thus the dissociation is inhibited. Also, for this reason, when $|m|$ is small, the configuration with the molecular axis parallel to the surface is more favorable for vibrational assisted adsorption, and thus the probability increases with the increase of ν.

3.4.3 Relation of Adsorption and Desorption Processes

In this section, the aim is to establish the relation between the adsorption and desorption processes. The processes where molecules scatter from the surface can be divided into two: direct and indirect scattering processes. The direct scattering process occurs when the impinging molecule is reflected due to the potential barrier. On the other hand, the indirect scattering involves a molecule that first adsorbs on the surface and then later desorbs. In the case where the surface can be approximated as a rigid body, both adsorption and desorption processes need to follow the same pathway, but in the reverse direction of each other [25–27].

Let E_t and E_ξ be the translational energy in the direction normal to the surface, and the internal energy of the molecule, respectively. From these definitions, the total desorption probability with the final state $\left(E'_t, E'_\xi\right)$ due to the initial state $\left(E_t, E_\xi\right)$ can be formulated as

$$\begin{aligned} P^D\left(E'_t, E'_\xi; E_t, E_\xi\right) &= P^D_{\text{direct}}\left(E'_t, E'_\xi; E_t, E_\xi\right) \\ &+ P^D_{\text{indirect}}\left(E'_t, E'_\xi; E_t, E_\xi\right), \end{aligned} \tag{3.60}$$

where P^D is the total desorption probability, P^D_{direct} is the desorption probability due to direct scattering process, and P^D_{indirect} is the desorption probability due to indirect scattering process.

Consider the desorption due to the direct scattering process P^D_{direct}. The desorption probability due to direct scattering process is assumed to obey the following master equation (Markovian approximation [28])

$$\begin{aligned} P^D_{\text{direct}}\left(E'_t, E'_\xi; E_t, E_\xi\right) &= q\left(E'_t, E'_\xi; E_t, E_\xi\right) \\ &+ \int dE''_t \sum_{E''_\xi} P^D_{\text{direct}}\left(E'_t, E'_\xi; E''_t, E''_\xi\right) q\left(E''_t, E''_\xi; E_t, E_\xi\right). \end{aligned} \tag{3.61}$$

The above equation can be interpreted as follows. The desorption probability due to direct scattering P^D_{direct} has two contributions. The first contribution, $q\left(E'_t, E'_\xi; E_t, E_\xi\right)$, is the transition probability from the initial state $\left(E_t, E_\xi\right)$ to the final state (E'_t, E'_ξ). The second contribution comes from the initial state (E_t, E_ξ) that transits to the intermediate states (E''_t, E''_ξ), before it transits again to the final state (E'_t, E'_ξ). Notice that the above equation is a recurrence equation, in which P^D_{direct} on the left-hand side can be substituted into the P^D_{direct} in the right-hand side (after properly adjusting the appropriate variables (E_t, E_ξ), which can be inferred that there can be multiple processes of (E''_t, E''_ξ), (E'''_t, E'''_ξ), and so on, before arriving at the final state (E'_t, E'_ξ)).

Based on the principle of detailed balance, the transition probability q must satisfy the following relation

$$q(E'_t, E'_\xi; E_t, E_\xi)e^{-(E_t,E_\xi)/k_B T_S} = (E_t, E_\xi; E'_t, E'_\xi)e^{-(E'_t,E'_\xi)/k_B T_S}, \tag{3.62}$$

where k_B is Boltzmann constant, and T_S is the temperature of the surface. In addition, the transition probability q has to be normalized as follows

$$\int dE'_t \sum_{E'_\xi} q(E'_t, E'_\xi; E_t, E_\xi) = 1. \tag{3.63}$$

Due to the detailed balance from (3.62), it can be seen that P^D_{direct} from (3.61) also satisfies the detailed balance relation.

Next, consider the desorption due to the indirect scattering process P^D_{indirect}. As mentioned earlier, the desorption from indirect scattering process occurs due to the desorption of the adsorbates that first adsorb on the surface. Let $S(E_t, E_\xi)$ be the sticking probality (which is simply another name of the adsorption probability) due to the impinged molecule with the initial state (E_t, E_ξ). Due to the conservation of probability, the probability in the direction of scattering molecules and the adsorption molecule (i.e., reflection and transmission probabilities) must sum to unity. Therefore, the sticking probability takes the form

$$S(E_t, E_\xi) = 1 - \int dE'_t \sum_{E'_\xi} P(E'_t, E'_\xi; E_t, E_\xi). \tag{3.64}$$

Consequently, the P^D_{indirect} depends on the sticking probability as follows

$$P^D_{\text{indirect}}(E'_t, E'_\xi; E_t, E_\xi) = D(E'_t, E'_\xi)S(E_t, E_\xi) \tag{3.65}$$

where D is the desorption probability from the adsorbates.

The next step is to relate the sticking probability S with the desorption probability from the adsorbed state D. At thermal equilibrium, both the incoming and scattering molecules at the surface temperature T_s are both described by Boltzmann distributions. Therefore, the following relation must hold

$$\int dE_t \sum_{E_\xi} \left[P^D_{\text{direct}}(E'_t, E'_\xi; E_t, E_\xi) + P^D_{\text{indirect}}(E'_t, E'_\xi; E_t, E_\xi)\right]$$
$$\frac{e^{-(E_t+E_\xi)/k_B T_S}}{k_B T_S Z(T_S)} = \frac{e^{-(E'_t+E'_\xi)/k_B T_S}}{k_B T_S Z(T_S)}, \tag{3.66}$$

where $Z(T_S)$ is the partition function for the internal states of the molecules. From the above relations, the desorption probability from the adsorbates D can be solved as in [25, 26]

$$D(E_t, E_\xi) = \frac{S(E_t, E_\xi)}{\langle S \rangle} \frac{e^{-(E_t+E_\xi)/k_B T_S}}{k_B T_S Z(T_S)}. \tag{3.67}$$

The quantity $\langle S \rangle$ has the following form

$$\langle S \rangle = \int dE_t \sum_{E_\xi} S(E_t, E_\xi) \frac{e^{-(E_t+E_\xi)/k_B T_S}}{k_B T_S Z(T_S)}. \tag{3.68}$$

From (3.67), the internal state distribution of $D(E_t, E_\xi)$ deviates from the Boltzmann distribution at temperature T_S if the sticking probability $S(E_t, E_\xi)$ depends on the internal state. Internal state cooling (heating) in desorption can be expected if $S(E_t, E_\xi)$ is a decreasing (increasing) function with respect to energy related to the internal state.

Equation (3.67) shows that when the conditions of the internal state ν and the translational kinetic energy E_t increases the adsorption probability, the desorption probability of the molecule at the same state also increases. It has been confirmed experimentally that this correlation is also strong [27]. As an example, the adsorption and desorption probabilities as function of energy are shown in Fig. 3.21. The

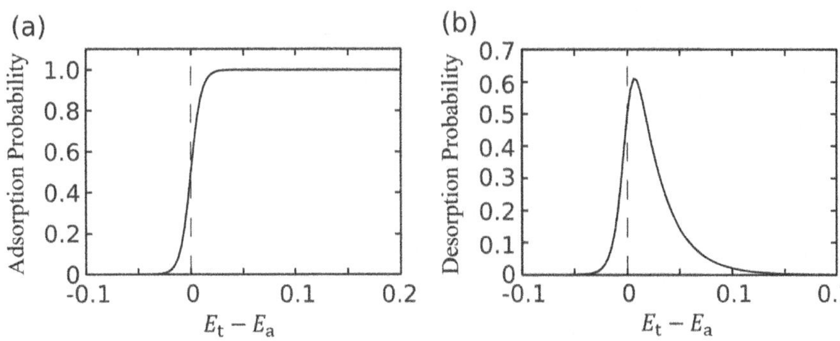

Fig. 3.21 Relationship between adsorption probability and desorption probability in the same surface system with activation barrier E_a. **a** Probability of dissociative adsorption of incident hydrogen molecules with translational kinetic energy E_t. **b** Probability distribution in which hydrogen molecules desorbed from the surface in thermal equilibrium have translational kinetic energy E_t. Adapted with permission from [55]. Copyright 2013 by Asakura Publishing

desorption probability concentrates in the vicinity of E_a because the energy dissipation during adsorption has a low probability in the presence of a high energy initial state in desorption. In addition, one can notice that the tunneling effect around the region E_a is remarkable.

3.4.4 Rotational Distribution in Associative Desorption

From the relation between adsorption and desorption discussed in Sect. 3.4.3, it can be observed that if the adsorption probability of molecules is independent of the distribution of the molecular internal states, orientations and velocities, equilibrium statistical mechanics predicts that the corresponding reverse process of desorption will be determined solely by the surface temperature. However, it is known from Sect. 3.4.2 that this is usually not the case; the PES of the molecule-surface systems are generally anisotropic to orientation (i.e., orientational sensitive). Such anisotropy effect to the desorbed molecule will be investigated using the rotational effect of the desorbed D_2/Cu(111) as an example.

Figure 3.22 shows the experimental and computational results for associative desorption probability $D(j, T_S)/(2j+1)$ from Cu(111) as a function of the final rotational energy E_j where j is the rotational quantum number. The desorption probability is the average for the desorption translational energy E_t and the magnetic quantum number m_j, and the molecular vibration is assumed to be in the ground

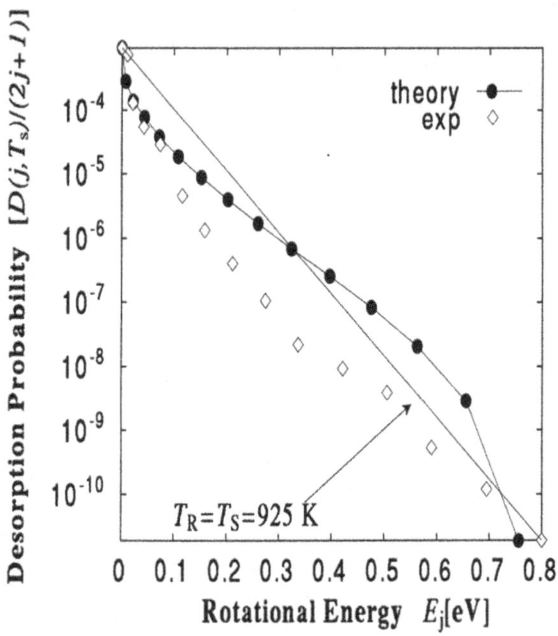

Fig. 3.22 Rotational energy E_j dependence of desorption probability in deuterium associative desorption from Cu(111). The circles and diamonds show the values obtained from quantum dynamics calculations and experiments, respectively. Each rotational energy corresponds to the rotational quantum number $j = 0, 1, \cdots, 14$. The straight line shows the Boltzmann distribution at a temperature of 925 K. Reprinted with permission from [22]. Copyright 2000 by Elsevier

state. For reference, the Boltzmann plot at $T_S = 925$ K is shown as the straight line in Fig. 3.22. If the rotational temperature T_R is equal to the surface temperature T_S, the desorption probability must show a Boltzmann distribution with respect to E_j, but a deviation from that distribution is found in experiments and calculations. The region of $j < 5$, where the steering effect was obtained by dissociative adsorption, falls in the region of $E_j < 0.1$ eV, and an increase in the probability (rotational heating) is observed in both experiments and calculations. The region of $5 < j < 10$ corresponds to the region of $0.1 < E_j < 0.4$ eV, and both the steering effect and the energy transfer between translation and rotation are small. Therefore, in the experiment, the probability is reduced (rotational cooling), and even though the calculation does not lead to cooling, suppression of heating is observed. Except for $j = 14$, the region of $j > 10$ corresponds to the region of $E_j = 0.4$ eV, and rotational heating is seen by the effect of energy transfer between translation and rotation.

Next, the phenomenon called quantum filtering will be discussed, which is related to the correlation between the final translational energy and the rotational alignment. In order to discuss the rotational alignment, the definition of a rotational orientation index $A_0^{(2)}$ (or quadrupole orientation factor) is utilized as follows [22, 29]

$$A_0^{(2)}(j, E_{\text{tot}}) = \frac{\sum_{m_j} \left[3m_j^2 - j(j+1)\right] D_{jm_j}(E_{\text{tot}})}{j(j+1) \sum_{m_j} D_{jm_j}(E_{\text{tot}})}. \tag{3.69}$$

Here, $D_{jm_j}(E_{\text{tot}})$ is the desorption probability of the molecule with the total energy $E_{\text{tot}} = E_t + E_j$. In the case of $A_0^{(2)} < 0$, the cartwheel-like rotation is dominant, while in the case of $A_0^{(2)} > 0$, the helicopter-like rotation is dominant. Figure 3.23 shows the schematic of the cartwheel-like rotation and helicopter-like rotation. The

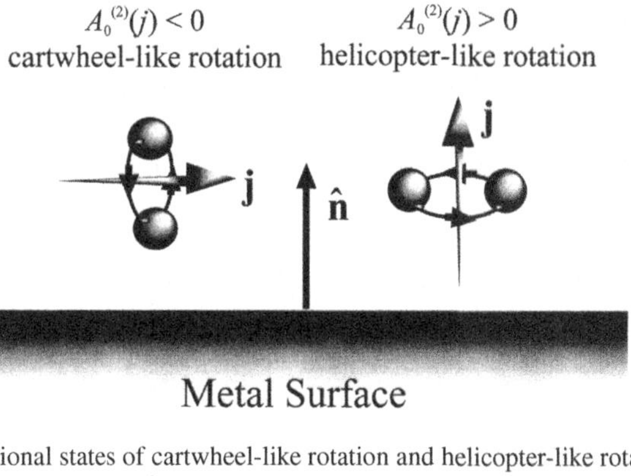

Fig. 3.23 The rotational states of cartwheel-like rotation and helicopter-like rotation with respect to the surface normal unit vector $\hat{n}$. The cartwheel-like rotation preference is associated with the orientational rotation index $A_0^{(2)}(j) < 0$, while the helicopter-like rotation preference is associated with $A_0^{(2)}(j) > 0$. Lastly, $A_0^{(2)}(j) = 0$ means the spatially isotropic distribution of j. Reprinted with permission from [22]. Copyright 2000 by Elsevier

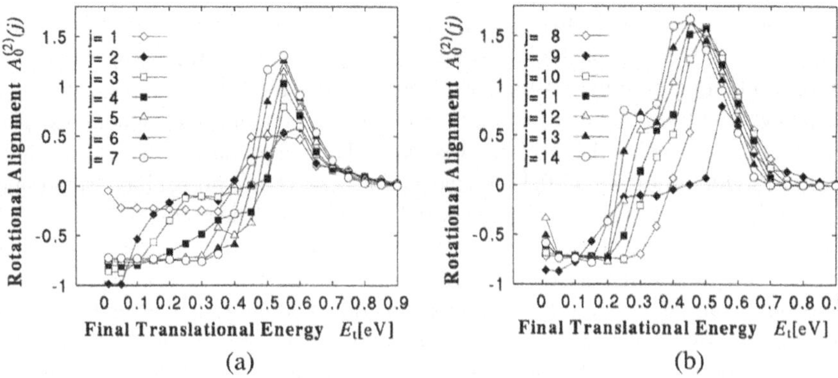

Fig. 3.24 Rotational orientation index $A_0^{(2)}(j)$ expressed as a function of desorption translational energy E_t in associative desorption of D_2 from Cu(111) [22, 29] (reprinted from the literature [29]). **a** When the rotational quantum number j is $1 \leq j \leq 7$. **b** When the rotating quantum number j is $8 \leq j \leq 14$. Reprinted with permission from [22]. Copyright 2000 by Elsevier

limit $j \to \infty$, $A_0^{(2)} \to 2$ represents a complete helicopter-like rotation. $A_0^{(2)} = 0$ represents a state in which the rotation axes are distributed isotropically.

Figure 3.24 shows the rotational alignment as the function of the final translational energy of the desorbing D_2 from Cu(111) surface. From the figure, the desorbing D_2 prefers cartwheel-like rotation at the low translational energy, regardless of the rotation state j. As the final translational energy increases further, the preference for the rotational alignment changes from cartwheel-like to helicopter-like. Finally, at relatively high final translational energy, the preference for the alignment converges to $A_0^{(2)} = 0$, signifying the isotropic orientational alignment.

To understand the above phenomena, consider the following argument. When the dissociatively adsorped atoms approach each other before crossing the activation barrier, they are most likely in a helicopter-like configuration because the two atoms are on a plane parallel to the surface. Since the desorbing molecule need translational energy that is at least around the same as the activation barrier, given the total energy E_{tot} of the molecule, the remaining energy for the rotational state of the molecule at the activation barrier is $E_{j,\text{crit}} = E_{\text{tot}} - E_{\text{barrier}}$. The $E_{j,\text{crit}}$ is the maximum rotational energy that the molecule could have for a given value of E_{tot}, and E_{barrier} is the activation barrier of the minimum energy path. Right after the molecule crosses the activation barrier, due to the anisotropy of the PES with respect to θ, the molecule can have either cartwheel-like rotation or helicopter-like rotation. It turns out that the anisotropy of the PES makes the cartwheel-like rotation have larger rotational energy than the critical rotational energy ($E_{j,\text{cart}} > E_{j,\text{crit}}$), while the helicopter-like rotation has the smaller rotational energy than the critical rotational energy ($E_{j,\text{helicop}} < E_{j,\text{crit}}$). As a result, if the E_{tot} of the molecule is high enough to cross the barrier but not extremely high, due to the law of energy conservation, the molecules with cartwheel-like rotation will have relatively smaller translation energy (because since $E_t = E_{\text{tot}} - E_j$ and E_{tot} is kept fixed, the high value of E_j in the case of cartwheel-like rotation yield the small value of E_t). Thus, the region of the small E_t is dominated with the rotational alignment less than zero ($A_0^{(2)} < 0$). Conversely, the helicopter-like rotation transfers some of the rotational energy to the translational energy, resulting in the helicopter-like rotation to dominate in the relatively high translation energy regime, yielding the positive rotational alignment ($A_0^{(2)} > 0$). Lastly, as the E_{tot} is very high, the energy difference of the cartwheel-like and helicopter-like rotations are relatively small as compared to the E_{tot}, and the cartwheel-like rotational state becomes accessible. Therefore, the rotational alignment converges to zero ($A_0^{(2)} = 0$) at the high energy regime.

3.4.5 Inelastic Scattering with Rotational Excitation

If the incident molecules on the surface cannot cross the activation barrier for dissociative adsorption, they are reflected and scattered toward the gas phase. In this scenario, inelastic scattering may occur in which the incident translational energy E_i is transferred to the internal energy of the molecule or surface. The waves of scattering molecules from different surface sites can also cause interference and diffraction. When only elastic scattering occurs, (i.e., Bragg reflection), the diffraction image gives a reciprocal lattice of the surface geometry. A diffraction spot whose translational wave number vector $\vec{K}_i$ in the surface parallel direction is changed to $\vec{K}_f$ by scattering is displayed as (m, n) in association with the change $\Delta\vec{G} = \vec{K}_f - \vec{K}_i = (mG, nG)$ of the reciprocal lattice vector, where $G = 2\pi/a$ is the reciprocal lattice constant for the lattice constant a. When inelastic scattering occurs, new spots may appear in the diffraction image, but they usually appear at the same diffraction spots as elastic scattering.

As an example, consider the inelastic scattering with rotational excitation on Cu(001). For hydrogen dissociative adsorption on Cu(001), there is an activation barrier of about 500–900 meV, and molecules that enter with lower kinetic energy cannot be dissociated and reflected to the gas phase side [30]. Steering is dominant in this kinetic energy region, and molecules that collide near the activation barrier are scattered while transferring energy and momentum from translational motion to rotational motion. Figure 3.25 shows the result of quantum dynamics calculation of the inelastic scattering probability that the rotating quantum number j changes from 0 to 2. However, this corresponds to the case where H_2 molecules that are oriented in the helicopter-like $m_j = 0$ and in the vibrational ground state $\nu = 0$ are incident at an oblique incidence angle along the [100] direction. The reflection depends on the orientation dependence of the PES and the corrugation of the PES in the parallel direction of the surface. Here, different internal degrees of freedom dependent on each diffraction spot. In particular, the conversion to the cartwheel-like is accompanied by a large change in momentum, which is represented by the diffraction spots [$(1, 0)$ and $(\bar{1}, 0)$]. A sudden change is seen at $E_i < 100$ meV, which reflects internal energy quantization (e.g., 76 meV for rotational excitation with $j = 0 \rightarrow 2$). Even when

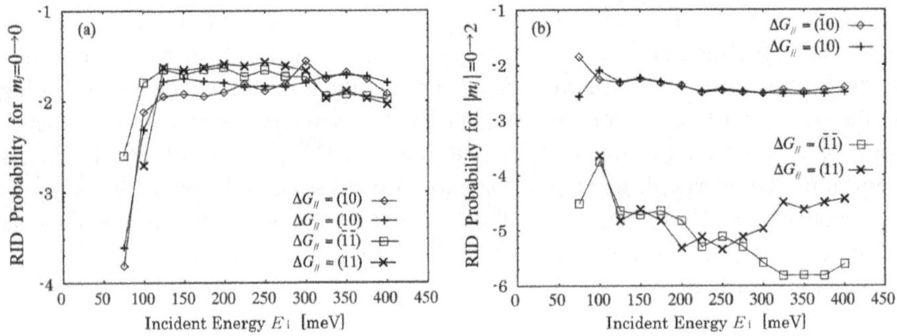

Fig. 3.25 Rotationally inelastic diffraction (RID) probability, in *logarithmic* plots, as the function of the incident translational energy E_i of hydrogen on Cu(001) changes from 0 to 2 and is inelastically scattered. The $\Delta G_{||}$ indicates a diffraction spot. **a** When the orientation after scattering is helicopter-like $m_j = 0$. **b** When the orientation after scattering is cartwheel-like $|m_j| = 2$. Reprinted with permission from [31]. Copyright 2000 by the Physical Society of Japan

$E_i > 100$ meV, the non-monotonic incident translational energy E_i dependence is seen due to the quantum nature of hydrogen motion, but globally, the higher the E_i (though it is more gradual), the more the steering is suppressed, the internal state changes tend to be smaller. Therefore, the probability of changing to the cartwheel-like is particularly low for high E_i. The above orientation dependence indicates that dynamic quantum filtering works in diffraction scattering.

Comparing between hydrogen H_2 and deuterium D_2, there is a large behavior difference due to isotope effect. Figure 3.26 shows the dependence of the inelastic

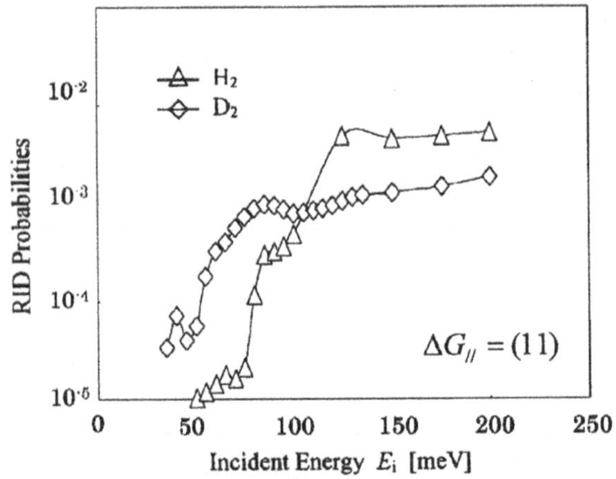

Fig. 3.26 Rotationally inelastic diffraction (RID) probability, in *logarithmic* plots, as the function of the translation energy E_i of hydrogen inelastic scattering probability on Cu(001). The diffraction spot is the same (1, 1) as in Fig. 3.25. The triangles and diamonds indicate values for hydrogen H_2 and deuterium D_2, respectively. Reprinted with permission from [32]. Copyright 2001 by Elsevier

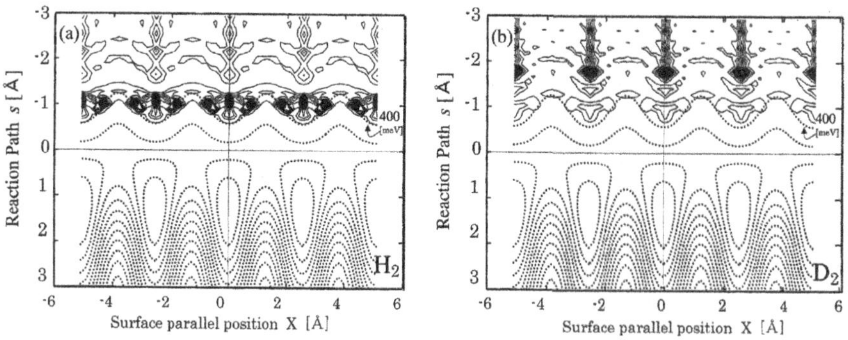

Fig. 3.27 Potential energy surface (dotted line) of **a** hydrogen H_2 and **b** deuterium D_2 on Cu(001) and contours of wave function (solid line) for $j = 0$, $\nu = 0$, $K_i = (0, 0)$ (reprinted from literature [33]). The horizontal axis is the coordinate X in the surface parallel direction of the molecule center, and the vertical axis is the reaction path coordinate s. The contour line spacing is 200 meV. Reprinted with permission from [33]. Copyright 2001 by the Vacuum Society of Japan

scattering probability with rotational excitation of $j = 0 \rightarrow 2$ on the incident translational energy E_i. Here, only the value at the diffraction spot (1, 1) is illustrated, but the qualitative difference between H_2 and D_2 is the same at other spots as well [32]. Since the PES for H_2 and D_2 are equivalent, the effect found is purely due to the difference in nuclear mass. For $E_i < 100$ meV, D_2 has a higher probability. This is because the internal energy quantization interval of D_2 is smaller than H_2 due to heavier nuclei of D_2 as compared to that of H_2, and therefore the excitation threshold is easier to reach. Conversely, H_2 has a higher probability when $E_i > 100$ meV. This is due to the quantum mechanical properties of nuclear motion near the activation barrier (Fig. 3.27).

Fig. 3.27 shows the nuclear wave function of a molecule incident at a kinetic energy of 200 meV with vibration and rotation as the ground state as a function of the surface parallel coordinate X and reaction path coordinate s. Here, the angle between the molecular axis and the surface normal is fixed to $\theta = 90°$, the azimuth angle of the molecular axis is fixed to $\phi = 0°$, and the surface parallel direction component perpendicular to X is fixed to $Y = 0$. Molecules flying from the gas phase side ($s < -3$ Å) collide with the activation barrier on the surface, but for both H_2 and D_2, the leaching of the wave function then exceeds the classical mechanical limit and the potential energy has reached the region of 400 meV. As a general rule of quantum mechanics, in the classically prohibited region, the wave function decays exponentially, and the wavelength for the same energy is inversely proportional to the square root of the mass. Therefore, the wavefunction in the case of impinged H_2 can penetrate to a deeper region than D_2, and the peak of the wave function is located in a region about 1 Å deeper than D_2 in the case of H_2. In other words, H_2 gets closer to the surface and is scattered in a region with larger orientation dependence due to the corrugation of PES. As a result, H_2 shows a higher inelastic scattering probability.

3.4.6 *Reactive Scattering with Stripping of Adsorbed Hydrogen Atom*

When a hydrogen atom beam or plasma is irradiated onto a surface on which a hydrogen atom is already adsorbed, a peeling reaction can occur in which the hydrogen atoms incident on the surface combine with the adsorbed hydrogen atoms and leave the surface. This microscopic reaction mechanism can be clarified quantitatively by quantum dynamics calculation based on the coupled channel method. When hydrogen atoms enter the surface where hydrogen is dissociated and adsorbed, it is first accelerated by the attractive force from the surface, and when it reaches the surface, it is scattered by periodic corrugation of PES. As a result, the motion of the impinged atom can change from surface normal direction to the parallel direction. The analysis based on the adiabatic potential energy can quantitatively evaluate the contribution of the Eley-Rideal mechanism and the hot atom mechanism (see Sect. 1.6.4).

Figure 3.28 shows the adiabatic potential energy surface for incident hydrogen atom motion in Cu(111) with hydrogen dissociated and adsorbed. However, the relative coordinate system of "molecules" composed of incident and adsorbed atoms is used here. When the incident hydrogen atom was impinged directly above the adsorbed atom ($X = 0.00$ Å, point A in Fig. 3.28), the interatomic distance r

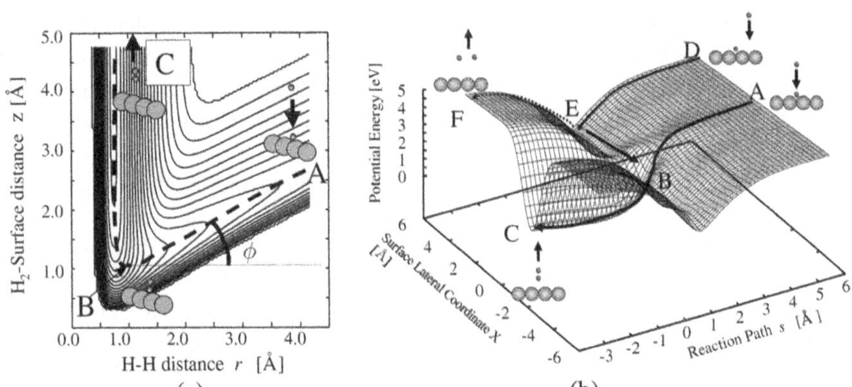

Fig. 3.28 Adiabatic potential energy surface [34–37] for incident hydrogen atom motion in hydrogen adsorbed on Cu(111). The surface parallel and vertical components of the center-of-mass coordinates of the molecules composed of the incident and adsorbed atoms are given by coordinate X and z (the origin of X is directly above the adsorbed atom), and the interatomic distance is given by r. Note that z is the mass-weighted coordinate as defined in the Fig. 3.15. **a** gives X directly above the adsorbed atom ($X = 0.00$ Å) and is expressed as a function of r and Z (reprinted from the literature [34]). The contour interval is 0.2 eV. The broken line shows the reaction path. **b** is expressed as a function of reaction path coordinates s and X. The curve with an arrow represents the reaction path when the incident X is away from the adsorbed atom. Reprinted with permission from [37]. Copyright 2002 by Elsevier

approaches the hydrogen molecule bond distance ($r = 0.74$ Å, point B). The interaction between the surfaces becomes repulsive, and the reaction proceeds in the direction of desorption as a molecule (point B to point C). This path corresponds to the Eley-Rideal mechanism. On the other hand, when hydrogen is impinged from a distance (point D) and collides with the surface, it first adsorbs on the surface and changes the direction of motion to parallel to the surface (point E), and after moving on the surface, it forms a molecule by collision with adsorbed atoms (point B), interaction from the surface becomes repulsive and leaves the surface (point C). This process corresponds to the process by the hot atom mechanism. When the incident kinetic energy is low, the reaction that follows this reaction path is likely to occur due to the steering effect.

However, if the incident kinetic energy is a little higher and the kinetic energy conversion in the parallel direction is larger, even if it approaches the adsorbed atom, it passes through the turning point (point B) before forming the molecule and remains trapped on the surface.

The mechanism of stripping reaction and scattering depends on the coverage of adsorbed hydrogen; the adsorbed hydrogen affects the periodic corrugation of PES. Since incident hydrogen atoms and desorbed hydrogen molecules behave as waves due to quantum mechanical effects, scattering due to corrugation of the PES causes diffraction phenomena. Figure 3.29 shows the H coverage Θ_H dependence of the angular distribution of the reaction probability $P_{reac}^{\Theta_H}(m_f, n_f = 0)$ that H_2 desorbs in the translational motion state m_f and the vibrational ground state $n_f = 0$. Here, the kinetic energy of the incident H atom is $E_t = 0.1$ eV and the incident angle is 10° (the surface parallel component of the kinetic energy is $E_i^{||} = 0.0035$ eV). The figure shows the resonance structures with peaks at specific angles can be seen for each coverage. This resonance is caused by quantum effects such as diffraction of incident H atoms and binding of H atoms on the surface. When incident H atoms collide with the surface, they change the direction of movement parallel to the surface by diffraction and diffuse while confined to a two-dimensional potential well on the surface.

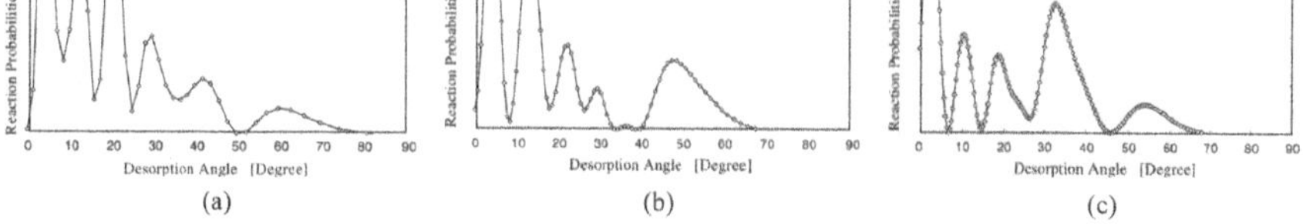

Fig. 3.29 Angular distribution of the probability that H_2 in the vibrational ground state is formed and desorbed by H atom incidence on H/Cu(111) (reproduced from [36]). Figures **a**, **b**, and **c** show the cases where the coverage ratio Θ_H of H is 0.500 ML, 0.250 ML, and 0.125 ML, respectively. The horizontal axis shows the H_2 desorption angle, and the vertical axis shows the reaction probability. The incident energy is $E_t = 0.1$ eV, and the incident angle is 10°. Reprinted with permission from [36]. Copyright by the Physical Society of Japan.

In this case, since kinetic energy transfer $\Delta E^{\perp}$ occurs from the surface normal component to the parallel component, the kinetic energy of the incident H atom is $E_f^{||} = E_i^{||} + \Delta E^{\perp}$ for the surface parallel component and $E_f^{\perp} = E_i - E_i^{||} - \Delta E^{\perp}$ for the vertical component. When it collides with adsorbed H atoms in the process of diffusion to form H_2, the surface repels the H_2 molecule, causing it to desorb from the surface. The ratio of $E_f^{||}$ and $E_f^{\perp}$ varies depending on the desorption angle of H_2, but at a specific angle, $E_f^{\perp}$ resonates with the bound level of the dimensional potential well, resulting in a peak reaction probability $P_{reac}^{\Theta_H}(m_f, n_f = 0)$. Here, the difference in coverage Θ_H only changes the corrugation of the PES in the direction parallel to the surface and does not affect the vertical direction, so the angle at which the peak appears hardly depends on Θ_H. The Θ_H dependence appears in the intensity of each peak.

The rate of energy transfer between degrees of freedom of movement reflects the type of main reaction mechanism. Here, the normalized average vibration energy

$$\left\langle E_{\mathrm{vib}}^{\Theta_H} \right\rangle = \sum_{n_f} E_{\mathrm{vib}}(n_f) \sum_{m_f} P_{\mathrm{reac}}^{\Theta_H}(m_f, n_f) / P_{\mathrm{tot}}^{\Theta_H} \tag{3.70}$$

of the desorbed H_2 molecule and the average translation energy

$$\left\langle E_{\mathrm{parallel}}^{\Theta_H} \right\rangle = \sum_{m_f} E_{\mathrm{parallel}}(m_f) \sum_{n_f} P_{\mathrm{reac}}^{\Theta_H}(m_f, n_f) / P_{\mathrm{tot}}^{\Theta_H} \tag{3.71}$$

in the surface parallel direction are defined. $P_{\mathrm{tot}}^{\Theta_H} = \sum_{n_f} \sum_{m_f} P_{\mathrm{reac}}^{\Theta_H}(m_f, n_f)$ is the total reaction probability, $E_{\mathrm{vib}}(n_f)$ is the vibrational energy for n_f, and $E_{\mathrm{parallel}}(m_f)$ is the translational kinetic energy surface parallel component for m_f. $\sum_{n_f}$ and $\sum_{m_f}$ are the sums for all n_f and m_f, respectively. Figure 3.30 shows the dependence of $\left\langle E_{\mathrm{vib}}^{\Theta_H} \right\rangle$ and $\left\langle E_{\mathrm{parallel}}^{\Theta_H} \right\rangle$ on the coverage Θ_H. When the coverage Θ_H is low, the chance of incident H atoms colliding directly with adsorbed H atoms is significantly reduced, so only the contribution of the hot atom mechanism remains. In the hot atom mechanism, the incident H atoms move in the direction parallel to the surface and collide with the adsorbed H atoms, and the probability of desorption with the two atoms moving in the direction parallel to the surface increases. Therefore, at low coverage Θ_H, $\left\langle E_{\mathrm{vib}}^{\Theta_H} \right\rangle$ and $\left\langle E_{\mathrm{parallel}}^{\Theta_H} \right\rangle$ have almost the same value. On the other hand, when Θ_H is high, the chance of the incident H atom colliding with the adsorbed H atom increases in both the Eley-Rideal mechanism and the hot atom mechanism. In the Eley-Rideal mechanism, incident H atoms move in the direction perpendicular to the surface and collide with adsorbed H atoms, so the direction of movement of the two atoms has a surface normal component. Therefore, the relationship of $\left\langle E_{\mathrm{vib}}^{\Theta_H} \right\rangle > \left\langle E_{\mathrm{parall}}^{\Theta_H} \right\rangle$ is found at high Θ_H.

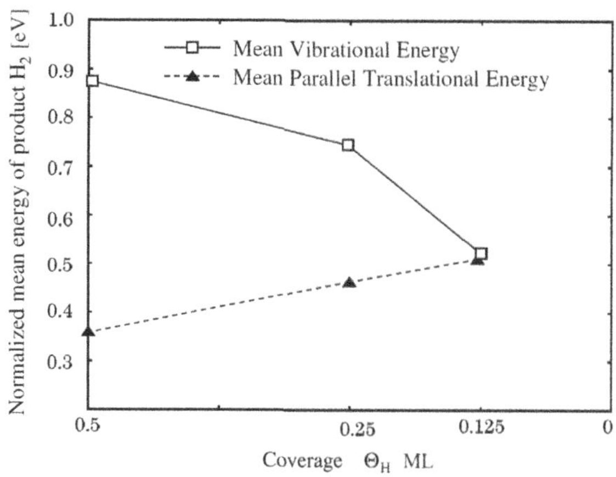

Fig. 3.30 Normalized mean vibration energy $E_{vib}^{\Theta_H}$ (white square) and mean parallel translational energy $E_{paral}^{\Theta_H}$ (black triangle) in the surface parallel direction on H coverage Θ_H when H_2 is formed and desorbed by H atom incidence on H/Cu(111). Reprinted with permission from [36]. Copyright 2002 by the Vacuum Society of Japan

3.5 Measurement Methods for Hydrogen Quantum Dynamics

In this section, a brief experimental measurement method of hydrogen quantum dynamics is be presented. Typical methods for measuring hydrogen motion on surfaces include low energy electron diffraction (LEED) [39, 40], high resolution electron energy loss spectroscopy (HREELS) [31, 42, 33, 44], Low-Temperature Scanning Tunneling Microscopy (LT-STM) [45], Resonance Nuclear Reaction Method (NRA) [46, 47].

LT-STM and LEED can observe the hydrogen adsorption structure in real space and reciprocal space, respectively. Although it is almost impossible to observe hydrogen by ordinary STM measurement, the spectrum for the STM image and the bias voltage can be measured by cooling the sample surface to a temperature region (100 K or less) where the kinetic energy of the hydrogen nucleus is sufficiently small. In the case of LEED, electrons with energy of several tens to several hundred eV are incident on the sample surface, and the interference pattern of the reflected electron wave is detected. Since electrons of this energy can only reach a depth of several nanometers from the surface, only information about the surface can be obtained. Therefore, the observed interference pattern is different from the usual diffraction method and corresponds to the reciprocal lattice of a two-dimensional lattice. Since hydrogen nuclei behave as quantum waves instead of classical particles, it is necessary to consider the influence of nuclear motion on electron motion in the analysis of the obtained results, which remains a challenge for simulation technology.

HREELS and NRA can measure the vibrational energy of hydrogen adsorbed on the surface and have found good agreement with the quantum mechanical calculations described in Sect. 3.1. In the case of HREELS, an electron with energy of several eV is incident on the sample surface, and the energy of the inelastically scattered electron is measured. The obtained energy spectrum gives information on energy loss due to vibrations on the surface and excitation of plasmons. From the dependence of the energy spectrum on the scattering angle, vibration modes in the direction perpendicular to the surface and in the parallel direction can be distinguished and detected. When an adsorbate is present on the surface, incident electrons are subjected to long-range Coulomb interaction from the electric dipole created by the adsorbate, and the scattering called dipole scattering can occur. In this case, since the electric field due to the electric dipole is shielded by electrons on the metal surface, the surface parallel component of the electric dipole cancels the electric charge mirror image, and thus only the interaction with the vertical component is detectable (surface vertical dipole selection rule). As a result, the loss spectrum in the specular reflection direction gives only the information due to the vibration mode in the direction perpendicular to the surface [48]. Table 3.1 shows the vibration excitation energy in the vertical direction of the surface by HREELS measurement compared with the quantum mechanical calculation [49].

Although it is difficult to verify the zero-point vibration state with HREELS, NRA can measure the wave function spread and zero-point vibration energy of hydrogen atoms adsorbed on the surface [46, 47, 38]. In this method, an ion beam collides with the sample surface to cause a nuclear resonance, and the generated particles such as ions, gamma rays, and neutrons are detected. By this method, the absolute value of hydrogen vibrational state below 0.1 atomic layer can be quantified. The hydrogen atoms adsorbed on the surface undergo zero-point vibration, and the resonance energy varies depending on the vibration direction of the hydrogen atom. For example, when the hydrogen atom is displaced in the same direction as the ion beam incident direction, the apparent frequency of the hydrogen atom is reduced by the Doppler effect. The resonance energy is higher than the resonance energy when the hydrogen atom is located at the center of the zero-point vibration. Conversely, when the hydrogen atom is displaced in the direction opposite to the ion beam incident direction, the resonance frequency decreases because the apparent frequency increases. Based on this principle, when the intensity of the nuclear reaction product is plotted as a function of the energy of the incident ion beam, the zero-point vibration energy is reflected in the resonance width (Fig. 3.31). This corresponds to observing

Table 3.1 Theoretical (Quantum mechanical) calculations of vibrational excitation energy (perpendicular to the surface) of hydrogen atom H and deuterium atom D adsorbed on Cu and Pd surfaces [49]. Experimental values are obtained by high-resolution electron energy loss spectroscopy (HREELS). The parentheses indicate the vibration excitation energy of the deuterium atoms, and the numbers in square brackets indicate the quantum numbers

	Theoretical value (meV)	Experimental value by HREELS (meV)
Cu(111)	135 [$n = 4$] [50]	129 [44]
Cu(001)	90.8 (67.2) [$n = 1$] [51]	69.5 (53.2) [40]
Cu(110)	108 (88) [$n = 7$] [51]	118 (80) [43]
Pd(111)	118 (91) [$n = 4$] [52]	124 [41]
Pd(001)	-	60.2 [42]
Pd(110)	103 (85) [$n = 9$] [54]	104 [42]

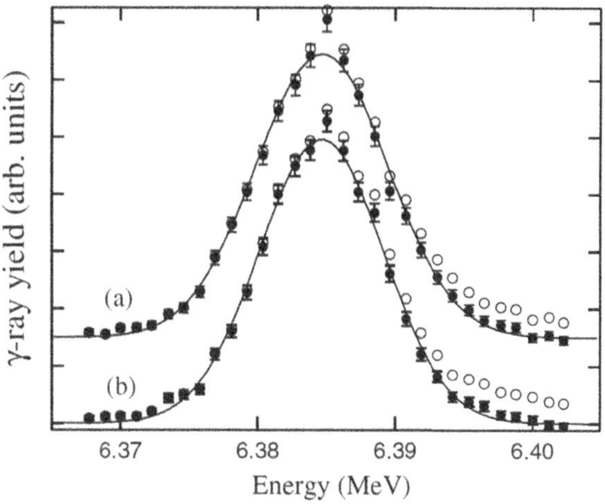

Fig. 3.31 Gamma line intensity spectrum measured by nuclear resonance method at H/Pt(111) (reproduced from literature [38]). The horizontal axis shows the energy of the incident ^{15}N ion beam. **a** and **b** show the incident angles of 0° and 45°, respectively. White and black points show the values before and after subtracting the background, respectively. Reprinted with permission from [38]. Copyright 2002 by the American Physical Society

the hydrogen wave function in momentum space. By scanning the ion beam incident angle, the vibration energy can be decomposed into a surface normal component and a parallel component for analysis. On the Pt(111) surface, it is known that the zero-point vibrational energy of hydrogen agrees well between NRA measurement and quantum state calculation [53].

References

1. K. Nobuhara, H. Nakanishi, H. Kasai, A. Okiji, Surf. Sci. **493**, 271–277 (2001)
2. H. Kasai, M. Tsuda, *Computational Materials Design, Case Study I: Intelligent/Directed Materials Design for Polymer Electrolyte Fuel Cells and Hydrogen Storage Applications (Japanese)* (Osaka University Press, Osaka, 2014)
3. E.F. Arguelles, H. Kasai, J. Vac. Sci. Technol. A **36**, 030601 (2018)
4. E.F. Arguelles, H. Kasai, J. Appl. Phys. **123**, 115301 (2018)
5. T. Sugimoto, K. Fukutani, Phys. Rev. Lett. **112**, 146101 (2014)
6. E. Bernard, C. Houriez, A.O. Mitrushchenkov, M. Guitou, G. Chambaud, J. Chem. Phys. **142**, 054703 (2015)
7. A. Farkas, *Orthohydrogen, Parahydrogen, and Heavy Hydrogen*, (Cambridge University Press, Cambridge, 1935)
8. E. Ilisca, Phys. Rev. Lett. **66**, 667–670 (1991)
9. E. Ilisca, Prog. Surf. Sci. **41**, 217–335 (1992)
10. K. Fukutani, T. Sugimoto, Prog. Surf. Sci. **88**, 279–348 (2013)
11. H. Kasai, M.C.S. Escaño (eds.), *Physics of Surface, Interface and Cluster Catalysis* (IOP Publishing, Bristol, 2016)
12. K. Pachucki, J. Komasa, Phys. Rev. A **77**, 030501(R) (2008)
13. H. Kasai, W.A. Diño, R. Muhida, Prog. Surf. Sci. **72**, 53–86 (2003)
14. R. Muhida, W.A. Diño, A. Fukui, Y. Miura, H. Kasai, H. Nakanishi, A. Okiji, K. Fukutani, T. Okano, J. Phys. Soc. Jpn. **70**, 3654–3659 (2001)
15. A. Groß, *Theoretical Surface Science* (Springer, Berlin, 2009)
16. T. Imamura, *Physics and Green's Function (Japanese)* (Iwanami Shoten, Tokyo, 1994)

17. Y. Takahashi, *Quantum Theory I for Condensed Matter Physics Researchers (Japanese)* (Baifukan, Tokyo, 1974)
18. W. Brenig, T. Brunner, A. GroB, R. Russ, Z. Phys. B. **93**, 91–101 (1993)
19. W. Brenig, R. Russ, Surf. Sci. **315**, 195–203 (1994)
20. W. Brenig, A. GroB, R. Russ, Z. Phys. B. **97**, 311–317 (1995)
21. H. Kasai, H. Akai, H. Yoshida (eds.), *Introduction to Computational Material Design (Japanese)* (Osaka University Press, Osaka, 2005)
22. W.A. Diño, H. Kasai, A. Okiji, Prog. Surf. Sci. **63**, 63–134 (2000)
23. W.A. Diño, H. Kasai, A. Okiji, J. Phys. Soc. Jpn. **64**, 2478–2487 (1995)
24. Y. Miura, H. Kasai, W.A. Diño, J. Phys. Soc. Jpn. **68**, 887–892 (1999)
25. H. Kasai, A. Okiji, Prog. Surf. Sci. **44**, 101–183 (1993)
26. H. Kasai, A. Okiji, J. Electron. Spectrosc. Relat. Phenom. **54**(55), 153–161 (1990)
27. M. Dürr, U. Höfer, Surf. Sci. Rep. **61**, 465–526 (2006)
28. H. Kasai, A. Okiji, Surf. Sci. **225**, L33–L38 (1990)
29. W.A. Diño, H. Kasai, A. Okiji, Surf. Sci. **418**, L39–L44 (1998)
30. G. Wiesenekker, G.J. Kroes, E.J. Baerends, J. Chem. Phys. **104**, 7344–7358 (1996)
31. Y. Miura, W.A. Diño, H. Kasai, A. Okiji, J. Phys. Soc. Jpn. **69**, 3878–3884 (2000)
32. Y. Miura, H. Kasai, W.A. Diño, A. Okiji, Surf. Sci. **493**, 298–304 (2001)
33. Y. Miura, H. Kasai, A. Okiji, J. Vac. Soc. Jpn. **44**, 276–279 (2001)
34. Y. Miura, H. Kasai, W.A. Diño, J. Phys. Condens. Matter **14**, 4345–4354 (2002)
35. Y. Miura, H. Kasai, A. Okiji, J. Vac. Soc. Jpn. **45**, 443–447 (2002)
36. Y. Miura, H. Kasai, W.A. Diño, A. Okiji, J. Phys. Soc. Jpn. **71**, 222–227 (2002)
37. Y. Miura, W.A. Diño, H. Kasai, A. Okiji, Surf. Sci. **507–510**, 838–844 (2002)
38. K. Fukutani, A. Itoh, M. Wilde, M. Matsumoto, Phys. Rev. Lett. **88**, 116101 (2002)
39. M. Rohwerder, C. Benndorf, Surf. Sci. 789–797 (1994)
40. I. Chorkendorff, P.B. Rasmussen, Surf. Sci. **248**, 35–44 (1991)
41. H. Conrad, M.E. Kordesch, W. Stenzel, M. Sunjic, J. Vac. Sci. Technol. A **5**, 452 (1987)
42. C. Nyberg, C.G. Tengstål, Phys. Rev. Lett. **50**, 1680–1683 (1983)
43. C. Astaldi, A. Bianco, S. Modesti, E. Tosatti, Phys. Rev. Lett. **68**, 90–93 (1992)
44. G. Lee, E.W. Plummer, Surf. Sci. **498**, 229–236 (2002)
45. L.J. Lauhon, W. Ho, Phys. Rev. Lett. **85**, 4566–4569 (2000)
46. M. Fukuoka, M. Okada, M. Matsumoto, S. Ogura, K. Fukutani, T. Kasai, Phys. Rev. B **75**, 235434 (2007)
47. M. Wilde, K. Fukutani, M. Naschitzki, H.-J. Freund, Phys. Rev. B **77**, 113412 (2008)
48. M. Nishijima, Hyōmen Kagaku (表面科学) **8**, 273–278 (1987)
49. N. Ozawa, M. Sakaue, H. Kasai, J. Vac. Soc. Jpn. **53**, 592–601 (2010)
50. K. Nobuhara, H. Kasai, H. Nakanishi, W.A. DIño, J. Appl. Phys. 5020–5025 (2004)
51. N. Ozawa, T. Roman, H. Nakanishi, W.A. Diño, H. Kasai, Phys. Rev. B **75**, 115421 (2007)
52. N. Ozawa, N.B. Arboleda Jr., T. Roman, H. Nakanishi, W.A. Diño, H. Kasai, J. Phys.: Condens. Matter **19**, 365214 (2007)
53. K. Fukutani, H. Kasai, Hyōmen Kagaku (表面科学) **27**, 213–219 (2006)
54. Allan Abraham B. Padama, Hiroshi Nakanishi, Hideaki Kasai, Quantum states of hydrogen atom on Pd(110) surface. Appl. Surface Sci. **359**, 687–691 (2015)
55. H. Kasai, M. Sakaue, *Surface and Interface Physics* (Asakura Publishing, Shinjuku, Tokyo, Japan, 2013)
56. B. Zwiebach, *8.06 Quantum Physics III* (Massachusetts Institute of Technology, MIT OpenCourseWare, 2018). https://ocw.mit.edu

Chapter 4
Review of the Current Status of the Hydrogen Economy

Abstract The envisioned hydrogen-powered future is gradually taking shape in our time. In many countries around the world, active efforts to decarbonize energy systems have turned hydrogen into the centerpiece of grand master plans for clean and sustainable replacement to fossil fuels. This chapter tackles the current status of hydrogen economy, which provides a scientific, social, and economic perspective of hydrogen economy and its realization in our modern time. It elaborates the importance of research findings on the different aspects—hydrogen production, storage, delivery, and fuel cell—relevant to the realization of a hydrogen economy. It also illustrates the impact of the results of scientific investigations in attaining innovative hydrogen technologies. Lastly, it discusses the current technologies for hydrogen in the context of their implementation following the hydrogen roadmaps of different countries around the world.

4.1 The Hydrogen Economy and Its Present Status

Hydrogen is one of the most important elements, which defines our existence, our survival, and our understanding of the physical universe. It is a vital component of living organisms because most of the molecules and the molecules essential for our survival, e.g. enzymes, proteins, water, food, and energy resources, are composed of H. Hydrogen is present from the beginning of time even before other elements, such as oxygen and gold, even existed [1]. It is believed that H is the mother of all elements, the substance from which other elements are made as stated in Prout's hypothesis [2]. Such hypothesis is supported by the idea that the atomic weights of other elements were integral multiples of that of the H. In science and engineering, the H atom, being a component of almost all molecules and compounds, makes it a requirement to first grasp the fundamental knowledge of its interaction with other materials for a better understanding of the more complicated systems.

The endless story of hydrogen is well documented by the available scientific publications from different branches of science, e.g. physics, chemistry biology, and others, discussing its fundamental aspects and/or elaborating its significance for

© Springer Nature Singapore Pte Ltd. 2020

H. Kasai et al., *Hydrogen and Hydrogen-Containing Molecules on Metal Surfaces*, Springer Series in Surface Sciences 71,
https://doi.org/10.1007/978-981-15-6994-4_4

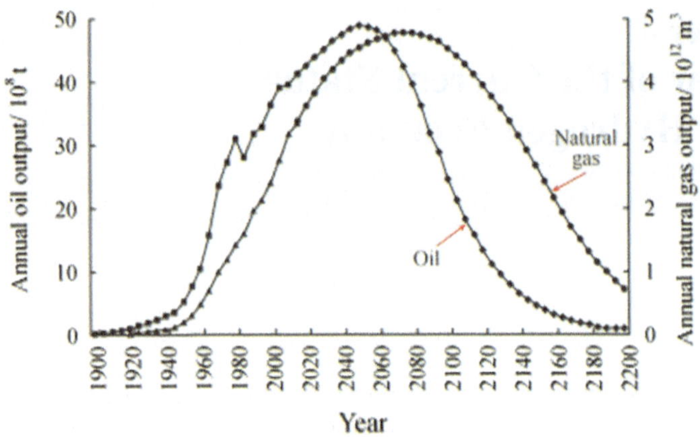

Fig. 4.1 Trends in the global production of oil and gas. Reprinted with permission from [4]. Copyright 2016 by Elsevier.

various applications. In the past decades, H was brought to limelight for its potential role as an energy carrier, which could solve our dependence on conventional fossil fuels to satisfy our energy demands. As an energy carrier, hydrogen is favored for its efficiency and for being environmental-friendly due to the fact that there is no emission of greenhouse gas in its utilization. In a typical hydrogen-based fuel cell, for instance, the efficiency (ratio of energy input and energy output) is reported to be 45–60% with only water and heat as the by-products [3]. On the other hand, conventional combustion of fossil fuels yields 33–35% efficiency and is accompanied by the emission of carbon dioxide and other pollutants that are detrimental to our environment. The presence of these pollutants in our atmosphere is considered as the major cause of global warming. In addition, fossil fuels are not renewable and are predicted to last a few decades from now. Figure 4.1 shows the forecast for oil and natural gas productions [4]. While there is an increasing trend in its current production, it is predicted that the peaks will be attained in 2040–2050. After that, the production will decrease which is definitely due to the scarcity in the available supplies. With the growing global energy consumption (see Fig. 4.2), due to the rapid increase in the population which translates to the increased demands for services, and to the economic growth of developing countries which requires buildings of infrastructures and others, it is a necessity to divert our attention to clean and renewable energy resources. Given these backgrounds, scientists and engineers continue to do their investigations to realize hydrogen-based technologies for energy-related applications.

Professor John O'.M. Bockris first used the term "Hydrogen Economy" in 1972 in a paper co-authored with Dr. John Appleby. This paper, published in Environment This Month journal, (1 (1972) 29) is titled "The Hydrogen Economy: An Ultimate Economy?" [6]. The paper discussed the potential use of hydrogen as a substitute to gasoline, and the associated cost of the technology necessary in its implementation. Thus, we can state that a *hydrogen-based economy* is a society in which the

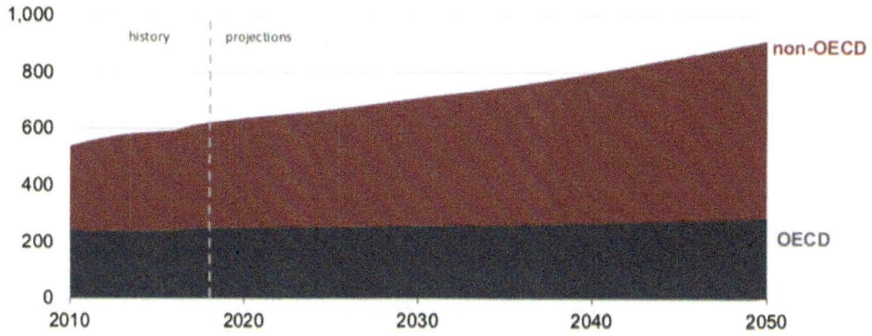

Fig. 4.2 History and projection of the world's energy consumption. OECD: organization of economic cooperation and development [5]

hydrogen is employed as the energy carrier in its energy system [7]. The complete integration of hydrogen in the energy sector will then eradicate the emission of CO_2 and other harmful pollutants from conventional fossil fuels. In other words, the hydrogen economy describes a society where hydrogen is the key player to satisfy its energy demands without harming the environment in its operation. In this section, the different aspects relevant to the realization of a hydrogen economy are discussed. The present status of the available technologies and the challenges in their implementations are also reviewed.

4.1.1 Source and Production

It is important to understand that hydrogen is an energy carrier, not an energy source. Thus, it should be produced. While it is desired to produce hydrogen from renewable and clean sources, fossil fuels remain as the main and the practical source of hydrogen gas. Nevertheless, it is still envisioned that clean and practical methods of producing hydrogen (with negligible emission of greenhouse gases) could be realized. Figure 4.3 illustrates the schematic diagram of a vision for hydrogen production from renewable resources in a hydrogen economy society [8]. In this vision, H is produced from water or other renewable sources. Such a process requires energy to activate the production or extraction of H. Renewable energy sources such as solar, wind, ocean, and biomass could provide this energy. The produced H is then utilized in industries, vehicles, and households. In the following sections, hydrogen production from renewable and non-renewable sources will be discussed.

4.1.1.1 Water Splitting: Solar Energy Assisted

In terms of sustainability and with consideration of minimizing the amount of carbon emission, water is the most promising source of hydrogen. A single H_2O molecule has two H atoms and one O atom. In principle, dissociating the molecule will yield H

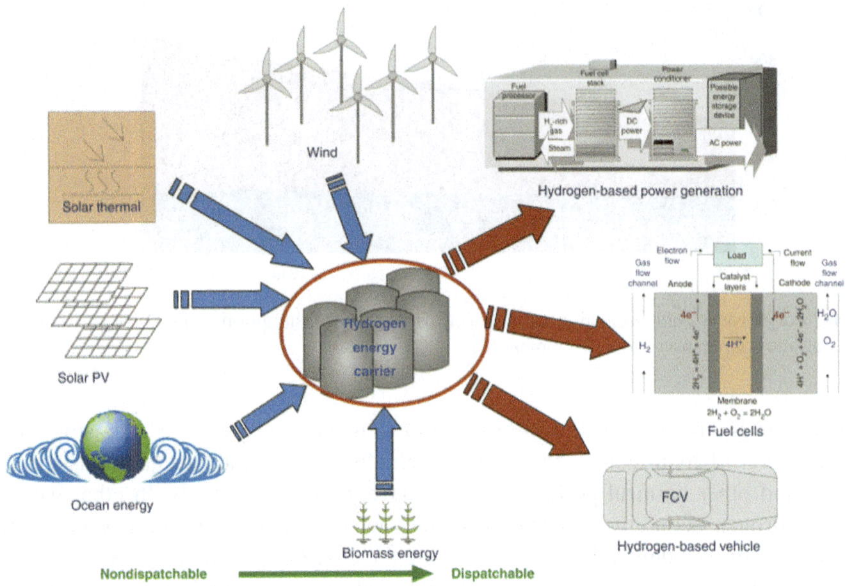

Fig. 4.3 Schematic diagram of a vision for hydrogen production from renewable resources in a hydrogen economy society. Reprinted with permission from [8]. Copyright 2016 by Elsevier

without worries of generating carbon byproducts. However, the atoms are strongly bonded with each other in the molecule. In the gas phase, it requires ~5.0 eV to dissociate the molecule into its component atoms [9]. Hence, catalysts and other external factors such as electrical energy or solar energy are necessary to assist the dissociation process.

Several technologies have been designed to split the water molecule, with the solar energy as external energy. Thermochemical water splitting is among these technologies, in which, so-called water-splitting cycles based on metal oxides are utilized. Solar energy is used to attain a significant amount of thermal energy which increases the temperature of the metal oxide. The material is reduced at high temperatures, which results in the formation of oxygen vacancies and the simultaneous release of oxygen species from it. Water reacts with the reduced material at low temperatures to produce hydrogen. Simultaneously, such reaction will oxidize the material and recycles it into its initial form [11, 12]. The schematic diagram of the process is shown in Fig. 4.4 [10].

The large solar concentrators necessary to attain the desired temperature for the system makes the process impractical for such application [13]. In addition, phase changes in the catalysts due to changes in temperature, cost of the operation and of the needed materials, and improvement and control of the kinetics of the reactions which could vary during the reduction and re-oxidation processes, are some of the challenges in this technology.

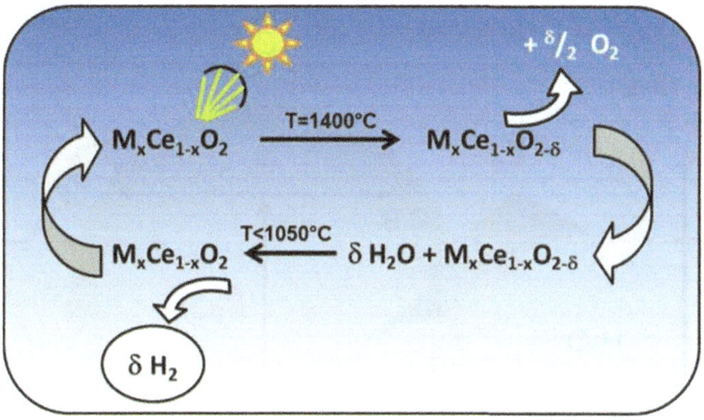

Fig. 4.4 Schematic diagram of water splitting via the thermochemical process with the use of metal oxides as catalysts. Reprinted with permission from [10]. Copyright 2012 by the American Chemical Society

Another technology is the photobiological water splitting. Here, microalgae and bacteria are used to split water, with the aid of sunlight, and produce hydrogen [14]. While hydrogen is realized through this process, increasing the yield of hydrogen and problems on scaling the process remain to be the crucial issues [13]. Other challenges include the possible inhibition of the hydrogen production by the produced oxygen during the splitting process and the activity and efficiency of the microorganisms in utilizing sunlight to produce more hydrogen gas.

The third technology based on solar energy is photocatalytic water splitting, in which, semiconductor catalysts are employed. A photon from sunlight excites the electron in the semiconductor. The electron will then activate the dissociation of the H_2O molecule. Due to the more advanced technology, photocatalytic water splitting is cheaper as compared to thermochemical and photobiological processes. Moreover, the hydrogen and oxygen gas stream could be easily separated in this technology which provides pure source of hydrogen gas [13]. The success of this technology greatly depends on the performance of the catalyst such that it could efficiently form electron–hole pair and prohibit its recombination. It is therefore important to consider or to match the bandgap of the semiconductor and the solar energy for a more efficient hydrogen production. Most of the materials being employed at present have large bandgaps. Thus, it is a challenge to improve the sunlight absorption ability of the material, via band gap engineering/modification, to efficiently utilize the visible light spectrum (which accounts for ~43% of the solar energy). Also, there are environmental issues on the use of semiconducting metal oxides as catalysts for such applications. The components of the degraded metal oxide could leak to the environment which may result in contamination [15]. Figure 4.5 shows the schematic illustration of the mechanism of water spitting using a photocatalyst

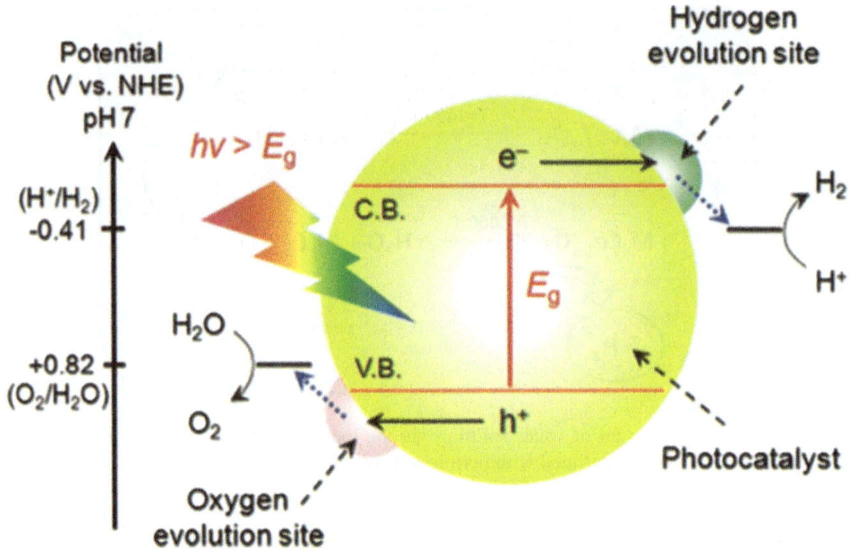

Fig. 4.5 Schematic diagram of the mechanism of water spitting using a photocatalyst. Reprinted with permission from [16]. Copyright 2010 by the American Chemical Society

4.1.1.2 Water Splitting: Electric Energy Assisted

Aside from solar energy, electric energy is also used to aid water splitting through the process known as electrolysis. In such technology, the process happens in an electrolyzer in which the system is designed similar to a fuel cell. In a typical polymer electrolyte membrane (PEM) electrolyzers, the water dissociates in the cathode side. The electrons (coming from the dissociated molecule) flow to an external circuit while the H^+ ions permeate through the membrane. The permeated H^+ combines with electrons to form H_2. On the other hand, the oxygen atoms from the dissociated molecule form oxygen gas. In this technology, an efficient electrocatalyst is necessary to effectively dissociate the water molecule. The schematic diagram of a PEM electrolyzer is shown in Fig. 4.6.

Most of the available PEM electrolyzers use iridium oxides as a catalyst. The material is stable in the acidic condition of the electrolytic cell. Iridium, however, is a precious metal and therefore, is not practical for such applications. Researchers have focused on reducing the amount of Ir by synthesizing Ir nanocatalysts to lower the cost of the system. Interestingly, recent discoveries found the efficient and stable performance of Mn oxides as a catalyst in the production of hydrogen via water electrolysis [18]. While Mn is abundant, previous works show that it degrades when used in PEM electrolyzers. However, the researchers found that the material will survive the acidic condition and will be stable (more than 8000 h of operation) if it is allowed to operate at a specific potential window. This potential window has been determined via spectroscopic measurements. Several works have likewise

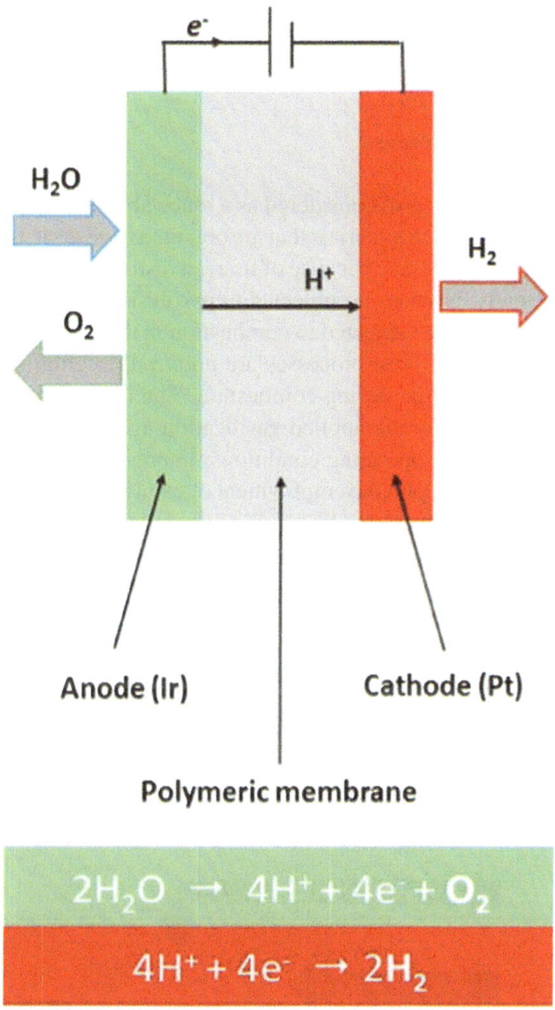

Fig. 4.6 Schematic diagram of a polymer electrolyte membrane (PEM) electrolyzer [17]

explored ways to improve the performance of other cheap transition metals for such applications.

There is increased activity in the scientific community to search for alternative materials for potential catalysts in water splitting applications. The cost of precious metals such as Pt, Ir and others, serves as motivation for such research endeavors. For example, Mo-based materials have been found to be possible materials for hydrogen evolution reaction [19, 20]. Other woks have also tested the performance of materials based on Co^{-}, Ni^{-}, and Fe [21, 22]. Certainly, this development shows a promising outcome and could possibly lead to cheaper electrolyzers in the future. Aside from reducing the cost of the catalysts, the efficiency of hydrogen production, stability in acidic and /or alkaline environment, and predicting the mechanism and rate-limiting process remain as issues in electrocatalytic water splitting.

4.1.1.3 Biomass

Biomass are also considered as a renewable hydrogen source. These are carbon- and hydrogen-rich materials that are obtained from agricultural and municipal wastes. It is a potential source because of its even distribution over the earth and of its abundance. Usually, biomass is subjected to pyrolysis and gasification processes, in which, the materials are subjected to combustion in the presence of small or negligible amounts of oxygen. These processes are preferred to eliminate the significant formation of carbon dioxide during combustion. The hydrogen production yield could be tuned depending on the applied gasification agent, e.g. steam [23], but will require high temperature operating conditions. Interestingly, lower temperature operation could be realized upon the employment of catalysts to assist the production via gasification [24]. This lower temperature signifies lower operating costs in the production process.

In principle, the combustion of biomass is not considered as a contributor to CO_2 emission because the plants take carbon dioxide in the atmosphere. The emitted CO_2 during combustion, therefore, is the amount taken while these plants are growing. Through gasification, a mixture of gas (CO, CO_2, H_2 and others) is produced. The schematic diagram of biomass gasification and production of the mixture of gas is shown in Fig. 4.7. Hydrogen production is further increased by allowing the interaction of CO with H_2O (the water gas shift reaction). While the technology already exists and is far matured than water splitting, the production of purified hydrogen gas, and simplifying the process for a more practical application limit this technology.

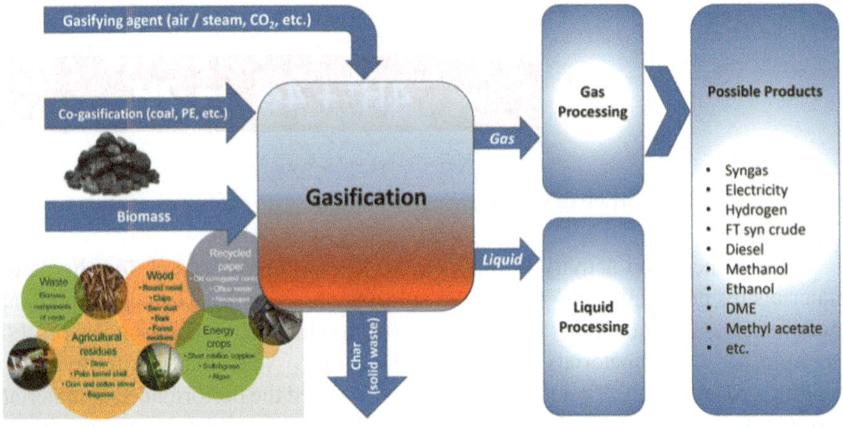

Fig. 4.7 Schematic diagram of biomass gasification and production of a mixture of gas. Reprinted with permission from [25]. Copyright 2016 by the BRTeam

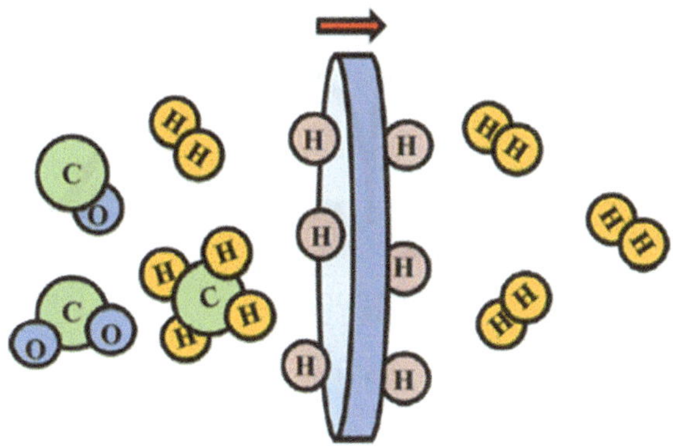

Fig. 4.8 Hydrogen separation using a membrane. Reprinted with permission from [26]. Copyright 2015 by Elsevier

The purification of hydrogen gas from biomass is usually done by using hydrogen separation and permeable membranes (Fig. 4.8). Most of these membranes are made from Pd, an expensive metal but is preferred for such application due to its reactivity toward hydrogen. H_2 molecule dissociates on Pd surfaces with a negligible amount of energy barrier [27]. This means that approaching H_2 is easily adsorbed as H atoms on the surface. Aside from this, Pd surfaces also facilitate the feasible absorption of H atoms [28–30]. This mechanism justifies experimental observations of the high ability of Pd based metals to absorb a large amount of hydrogen [31]. In spite of these properties of Pd toward hydrogen, a large amount of hydrogen in the bulk results in the expansion of the system which eventually embrittles the material [32]. In addition, Pd is prone to poisoning or is passivated by other molecules because of its reactive property in general. H_2 from biomass source is produced with other gases such as CO. The adsorption of CO blocks the active sites for H_2 dissociation and H atom adsorption and absorption [26]. A recent study also observed that the co-adsorption of CO and H on Pd surfaces can lead to CO hydrogenation which inhibits the interaction of the surface with hydrogen. These issues on the use of pure Pd are solved through alloying with other transition metals such as Cu, Au, and Ag [33–36]. Nonetheless, these alloyed systems still require large amount of Pd to retain the desired properties of the material for hydrogen separation applications.

Attempts to reduce the cost of the hydrogen separation membrane greatly depend on the amount of Pd in the membrane. It is interesting to note that among the tested metals, Group 5 elements were identified as materials with high hydrogen permeability, even higher compared with that of the Pd [37, 38]. This behavior is attributed to the body-centered cubic (bcc) structure of these elements which offers more favored interstitial sites for hydrogen diffusion as compared to the face-centered cubic (fcc) structure of Pd [39]. However, these metals cannot efficiently dissociate H_2 molecules; are more prone to embrittlement; and are easily oxidized. The cheaper cost of Group 5 materials than Pd leads to recent studies for the purpose of improving their mechanical stability and overall performance. The issues of embrittlement and mechanical stability are addressed by alloying the metals with other metals such as Pd, Ni and Ti [40]. The reported results, for V-Pd for instance [41, 42], depict

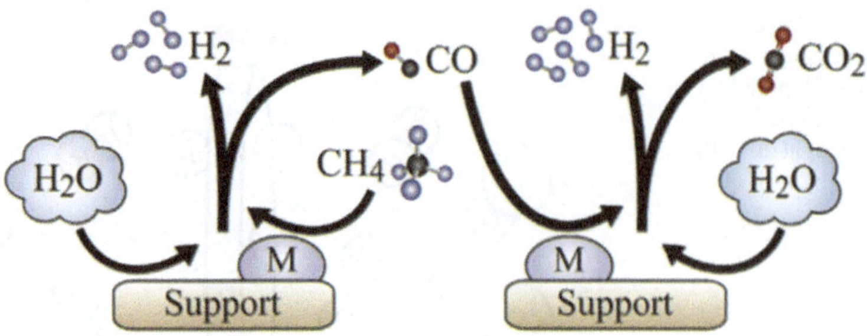

Fig. 4.9 Schematic diagram of the steam-methane reforming reaction and the subsequent water gas shift reaction for hydrogen production from methane. Reprinted with permission from [43]. Copyright 2014 by Elsevier

the improved performance of the alloyed materials specifically of its mechanical stability. These present conditions on gasification and hydrogen separation imply the important role of catalysts for a more practical and efficient reaction process.

4.1.1.4 Natural Gas: Steam Methane Reforming Process

At present, 95% of the hydrogen gas used in industries comes from natural gas due to its cheaper cost as compared to other processes. Natural gas is rich in methane (CH_4) and hydrogen is produced via the steam-methane reforming reactions. In such a case, methane interacts with water with the aid of a catalyst to produce hydrogen. The produced CO is again mixed with water to increase the amount of hydrogen produced. Alternatively, methane could also undergo partial oxidation to eliminate the CO_2 produced in the process. As such, only CO and H_2 are the byproducts, in which, water gas shift reaction could again increase the yield of H_2. The overall process of hydrogen production from methane with the aid of catalyst is shown in Fig. 4.9 [43].

Ni-based catalysts are usually employed for steam-methane reforming because of their catalytic performance and low cost compared to other metal catalysts [44]. The catalyst, however, suffers from passivation and deactivation due to the adsorption of C [45]. Alloying the Ni with other transition metals is one of the proposed solutions for such problem. In relation to the alloying of Ni, a recent study shows that substituting the subsurface atoms in Ni stepped surfaces with other elements will modify the reaction mechanism of methane decomposition [46]. This would likely prohibit the passivation of C on the surface through the weaker adsorption of intermediate species. More importantly, the performance of the alloy material is better than Ru, which is the usual catalyst employed in methane decomposition, for the purpose of eradicating coke formation (adsorbed-carbon passivation) on the surface.

One of the major drawbacks of natural gas as a source of hydrogen is the emission of a large amount of CO_2 [47]. If the emitted CO_2 is not captured or stored, then the reforming process could not be considered as an environmental-friendly process of producing H_2. CO_2 capture and storage (CCS) systems should be efficiently designed to retain a zero-emission of hydrogen production from these processes [48]. The captured molecules could be stored underground, be utilized in enhancing oil recovery or be converted into other useful compounds. Converting CO_2 into high-value chemicals (e.g. methanol, methane, etc.) could likewise aid in the production of fuels.

4.1.2 Storage and Delivery

One of the key elements in hydrogen energy systems is hydrogen storage. The technologies for storing hydrogen are required to work at various operating conditions and capacities to support the demand of consumers. Hydrogen storage technologies can be divided into two groups depending on their applications: *stationary* and *mobile* applications. Stationary hydrogen storage methods are mainly for on-site storage at either point of production or use, and for stationary power generation. On the other hand, mobile hydrogen storage applications are either for the purpose of transporting the stored hydrogen to point of storage or use, or consumption of hydrogen in a vehicle (e.g., automotive, aerospace, and maritime industry) [49]. Considering the low energy density by volume of hydrogen compared to fossil fuels, [50] extremely large storage vessels may be necessary to store a sufficient amount of hydrogen for large-scale use. To mitigate this issue, hydrogen storage technologies should provide high storage pressure, low storage temperature, or use materials with high hydrogen uptake.

Moradi and Groth have categorized hydrogen storage technologies as either *physical-based* or *material-based*, as demonstrated in Fig. 4.10 [49]. The physical-based hydrogen storage methods include storing hydrogen as compressed gas,

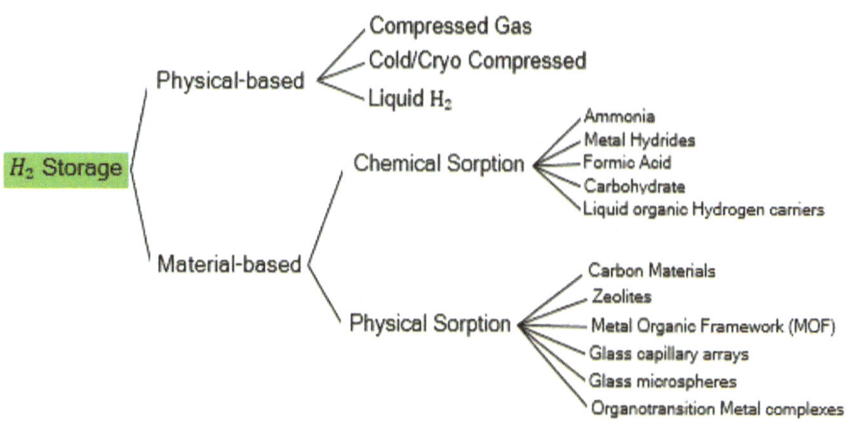

Fig. 4.10 The different methods for hydrogen storage. Reprinted with permission from [49]. Copyright 2019 by Elsevier

cold/cryo-compressed, and liquid hydrogen. On the other hand, material-based storage methods involve storing hydrogen by its chemical sorption (chemisorption) or physical sorption (physisorption) in a material.

A straightforward physical-based method of hydrogen storage is by using pressure vessels that store hydrogen as compressed gas. Moradi and Groth categorized these vessels as the following:

- **Type I: Fully metallic pressure vessels**. These are typically made from aluminum or steel and is considered to be the most conventional, least expensive, and heaviest. These pressure vessels weight approximately 3.0 lb/L and can withstand pressures up to 50 MPa.
- **Type II: Steel pressure vessel with a glass fiber composite overwrap**. This pressure vessel has the highest pressure tolerance and weights about 30–40% less than that of the Type I vessel. However, it is more expensive, costing about 50% more than Type I.
- **Type III: Fully composite wrap with metal liner**. This vessel offers about half less weight than Type II vessels and has proven to be reliable for 45 MPa working pressure. However, the cost is about twice that of Type II vessels. For Type III vessels, the structural load is mainly carried by the composite structure (carbon fiber composite), while the liner (aluminum) is used for sealing purposes.
- **Type IV: Fully composite**. This pressure vessel is the lightest and can withstand pressures of up to 100 MPa, but the price is relatively very high. The structural load is carried by carbon fiber or carbon-glass composites, with polymers such as High Densty Polyethylene for the liner.

Another physical-based method to storing hydrogen is by liquifying it at very low temperature (about −250 °C). However, this Liquid/Cryogenic hydrogen storage technology is challenged by the difficulty in maintaining hydrogen at such low temperature. Moreover, the process of hydrogen liquification is both time and energy consuming, with an energy content loss of up to 40% in the liquification process, compared to only about 10% energy loss in the compressed hydrogen storage methods. The liquid hydrogen storage method can usually be used for medium to large-scale storage and delivery such as in truck delivery and intercontinental hydrogen shipping. A cryogenic tanker can carry about 5000 kg of hydrogen, which is about five times the capacity of compressed hydrogen gas tube trailers.

Compressing hydrogen gas at about −233 °C is another physical-based hydrogen storage method. A cryo-compressed hydrogen is a super critical cryogenic gas, which does not liquify. This method of hydrogen storage promises high level of storage and safety as it provides about 80 g/L storage density, quick and efficient refueling, and a vacuum enclosure that ensures safety. In a comprehensive technical assessment of Ahluwalia et al., it was concluded that the cryo-compressed hydrogen storage method has the potential of meeting the requirement of the US Department of Energy for system gravimetric capacity, system volumetric capacity, and hydrogen loss during dormancy [51].

For material-based hydrogen storage, hydrogen is stored by its chemisorption or physisorption in the materials, as mentioned earlier. In chemisorption, the hydrogen

Table 4.1 Maximum storage capacities (%wt) reported for a number of different chemical and physical hydrogen storage methods. Reprinted with permission from [49]. Copyright 2019 by Elsevier

	Materia-based storage method	Maximum reported storage capacity (%wt)
Chemical	Ammonia Borane	19.4
	Metal Hydrides	12.6
	Alanates	9.3
	Formic Acid	4.4
	Carbohydrate	14.8
	Liquid Organic Hydrogen Carriers	7.2
Physical	Carbon Materials	8
	Zeolites	9.2
	Glass Capillary Arrays	10
	Glass Microspheres	14

molecules dissociate and integrate with the chemical structure of the material. The most popular materials for this purpose are metal hydrides, which offer a storage capacity of about 12.6%wt. On the other hand, hydrogen storage by physisorption uses porous materials to achieve high capacity and reliable storage units. The most promising porous materials for this purpose are the Metal Organic Framework (MOFs), and the porous carbon materials. A comparison of different material-based storage methods and their corresponding maximum reported storage capacities are shown in Table 4.1.

Aside from the storage of hydrogen, another major concern in hydrogen energy systems is the delivery of hydrogen to the end users. Depending on the storage method and application, hydrogen delivery can be through gaseous or liquid delivery methods, or through material-based hydrogen carriers. The choice of the delivery method will largely depend on factors such as geography, market and consuming behaviors, population density, size of refueling stations, among others.

Gaseous hydrogen can be transported with pipelines or tube trailers. In the US, hydrogen delivery via pipelines will require hundreds of thousands of kilometers of dedicated hydrogen pipelines. This massive infrastructure requirement necessitates exploring the possibility of using the current natural gas pipeline infrastructure to distribute hydrogen across the country. On the other hand, utilizing tube trailers for hydrogen delivery is generally the simplest method in terms of infrastructure requirements, with minimal hydrogen loss and low compression cost at fueling stations.

Compared to gaseous hydrogen delivery, liquid hydrogen delivery is considered to be economical for high demands and mid range distances, and is anticipated to be a major means of delivery in hydrogen fueling stations by 2020–2050 [52]. There are eight existing liquefaction plants in North America with production capacity of 5–10 metric tons per day [49]. To supply a sufficient amount of hydrogen in the future via the liquid hydrogen delivery method, it is required to have more liquefaction plants with higher production rates, less specific energy consumption, lower capital cost, and higher efficiency.

For low demand applications, material-based hydrogen delivery can be appropriate. Material-based storage and delivery offers higher safety levels due to low storage pressure, manageable properties at ambient conditions, and good gravimetric density.

4.1.3 Fuel Cells

Hydrogen fuel cells for vehicular applications are among the most advanced area of research when one talks about hydrogen-related technologies. Fuel cells, in general, are devices that convert the chemical energy of substances into electric energy. There are various kinds of substances that can be used as energy carrier and / or fuels in fuel cells such as natural gas, methanol, ethanol, and hydrocarbons. It is an advantage that fuel cells do not require combustion in its operation, thus, it eliminates or reduces airborne pollutants [53]. When pure hydrogen is used as the energy carrier in fuel cells, the only output is heat and vapor without any emission of greenhouse gases.

Figure 4.11 shows the different types of fuel cells and their properties. This classification is based on the type of electrolyte used. It can also be classified based on the fuel (methanol, borohydride, hydrazine, hydrogen, etc.) employed. A fuel cell is composed of an electrolyte sandwiched by electrodes (the anode and the cathode). In

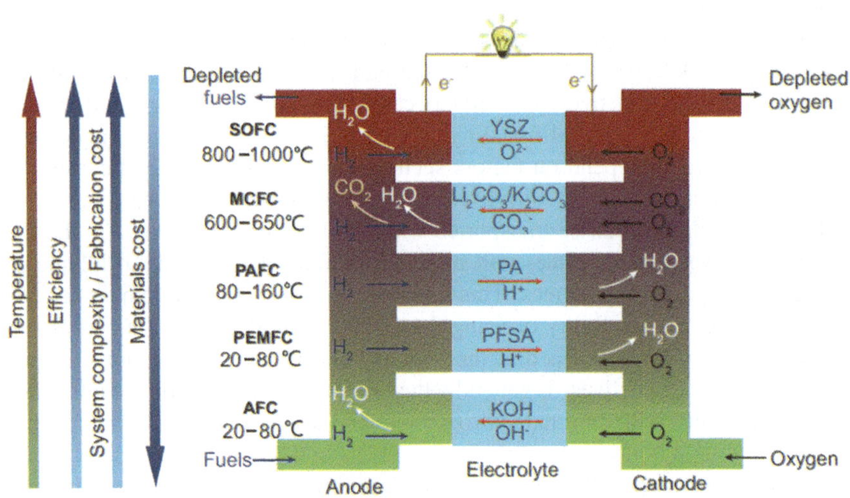

Fig. 4.11 Different types of fuel cell (classified based on the type of electrolyte) and their properties: alkaline fuel cell (AFC), proton-exchange membrane fuel cell (PEMFC), phosphoric acid fuel cells (PAFC), molten carbonate fuel cell (MCFC), and solid oxide fuel cell (SOFC). Reprinted with permission from [54]. Copyright 2017 by Oxford University Press

general, fuel cells operating at higher temperatures have higher efficiency and lower materials cost. Such systems, however, are usually accompanied by more complex designs and more expensive fabrication costs. On the other hand, proton-exchange membrane fuel cell (PEMFC) and alkaline fuel cell (AFC) are the most suited for low-temperature applications such as in fuel cell vehicles. These fuel cells differ in the electrolytes used in the cell, and consequently, in the types of materials employed.

For PEMFCs, the electrolyte is acidic and must allow the permeation of proton (H^+) through it. Perfluorosulfonic acid (PFSA)-based membranes such as Nafion® are the most advanced type of membranes that facilitate fast proton transfer. Due to the acidic nature of the electrolyte, PEMFCs require durable electrodes. Precious metals, such as the Pt group metals, are the most efficient electrodes that can withstand the acidic environment of the electrolyte. Thus, PEMFCs are usually challenged by the high cost of the electrode catalysts. On the other hand, AFCs use alkaline electrolyte which is desired to let the flow of anions through it. Non-precious based catalysts could be employed in AFCs which is an advantage in terms of practicality as compared to PEMFCs. However, development in AFCs is hindered by the slow anion conduction in the electrolyte layer. Further investigations are necessary to overcome this issue.

The performance of fuel cells strongly depends on the efficiency of the catalysts. In the case in which H_2 is used in fuel cells, the molecule undergoes chemical reactions and is converted to other chemical species while simultaneously providing electricity and heat. The law of conservation of energy dictates that for a given unit of the chemical reaction (such as $\frac{1}{2}\,O_2 + H_2 \rightarrow H_2O$ for the PEMFCs), the total energy output in the form of electrical energy and heat should be constant. Therefore, the desirable catalysts should be able to maximize the fraction of the electrical energy output (hence, automatically minimize the heat output). Also, the catalysts should have high current density. The current density depends on how fast the reaction can proceed, and thus, the catalysts should allow fast reaction of the needed processes.

Among the various types of fuel cells, the PEMFCs are the most investigated and understood. Its operation is shown in Fig. 4.12. Hydrogen and oxygen gas are dosed in the anode and cathode catalysts, respectively. Adsorption and dissociation of the gas molecules happen with the aid of the catalyst. When H_2 dissociates on the anode, its electrons diffuse resulting to current. The H^+ permeates through the electrolyte and combines with O and with the electron to form H_2O molecules. As shown in the figure, the only output is heat and vapor without any emission of greenhouse gases. The bottleneck reaction in PEMFCs is the oxygen reduction reaction (ORR) that occurs in the cathode. The interaction of the O_2 molecule is too slow relative to the interaction of H_2 in the anode.

Given the challenges in ORR in fuel cells, numerous attempts to understand such reaction and to aid in the design of appropriate catalysts have been done in the past. Among these approaches apply the Sabatier principle for catalyst design which states that the desired catalysts must have *appropriate* descriptors. Appropriate means that the values associated with the descriptors must not be too large nor too small. In such cases, theoretical designs were established with the adsorption of intermediate species such as oxygen, hydroxyl, and peroxyl as the descriptors for the reactivity of the desired catalysts. Figure 4.13a demonstrates the Sabatier principle which shows

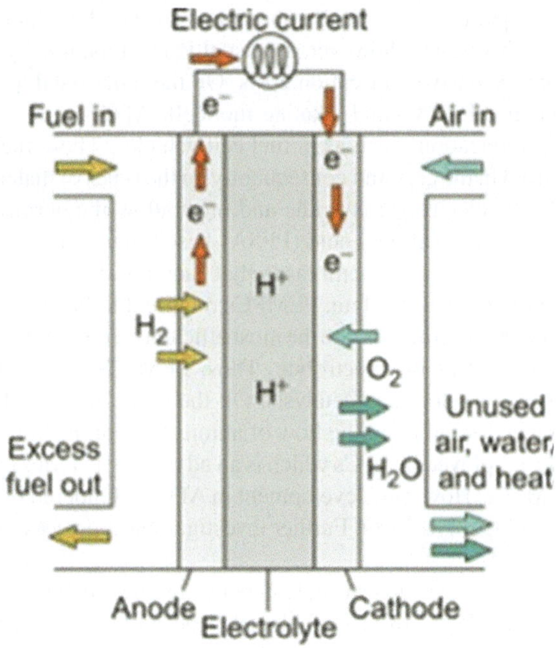

Fig. 4.12 Schematic diagram of a proton exchange membrane fuel cell (PEMFC). Reprinted with permission from [55]. Copyright 2018 by Elsevier

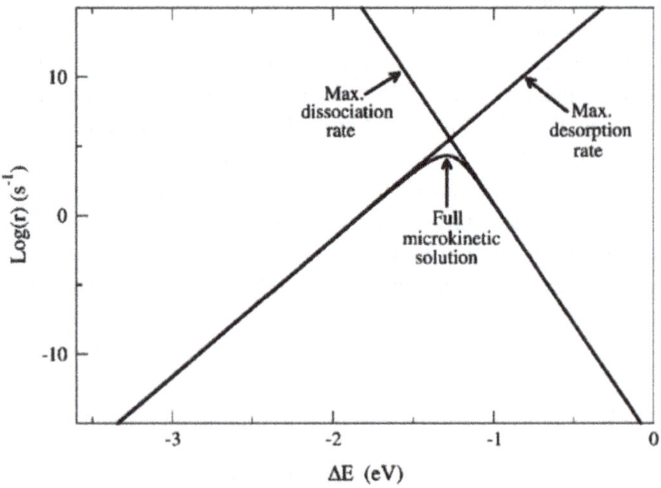

Fig. 4.13 Sabatier principle for catalyst design: volcano plot showing the relationship between adsorption energy and rate of desorption. Reprinted from [56]. Copyright 2008 by The Royal Society of Chemistry

		ICE	FCEV	BEV
Lower is better	Current capital cost	$	$ $ $	$ $
	Fuel cost	$ $	$ $ $	$
	Maintenance costs	$ $ $	$	$
	Infrastructure needs	$	$ $ $	$ $
	Emissions	♦ ♦ ♦	♦	♦
Higher is better	Efficiency	*	* *	* * *
	Range	* * *	* * *	*
	Refuelling speed	* * *	* * *	*
	Lifetime	* * *	* * *	* *
	Acceleration	* *	* * *	* * *

(a)

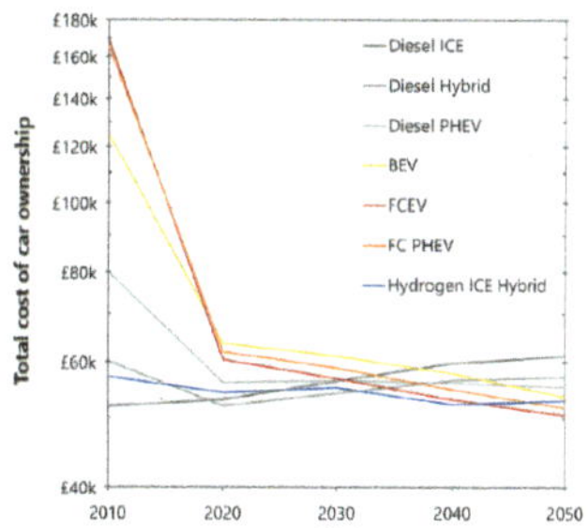

(b)

Fig. 4.14 **a** Costs and performance comparison among internal combustion engines (ICE), fuel cells electric vehicles (FCEV), and battery electric vehicles (BEV). Symbols give the qualitative comparison among different properties [57]. **b** The total cost of ownership for major powertrains. Reprinted with permission from [58]. Copyright 2014 by Elsevier

the relationship between adsorption strength and rate of desorption. Molecules with large adsorption energy have high dissociation rate. However, the strong adsorption hinders the species to desorb from the surface which negatively affects the catalytic activity. On the other hand, if the adsorption is too weak, the molecules do not adsorb and dissociate on the surface. Such "volcano plot", therefore, predicts that the material with adsorption energy corresponding to the peak of the volcano will have the best activity. That is, the catalyst that allows moderate adsorption strength of the adsorbate will provide the best activity toward the desired reaction. It is therefore important to design and tune the catalyst to obtain appropriate adsorption energies of the intermediate species and maximize the reaction rate.

In these past years, employment of fuel cells for vehicular applications have been started. Despite the convenience of internal combustion engines (ICE) and of the emergence of battery electric vehicles (BEV), researchers and engineers remained motivated of optimizing the performance and cost of fuel cells electric vehicles (FCEV). Figure 4.14a shows the costs and performance comparison among ICE, FCEV, and BEV. It is shown that FCEVs require minimal maintenance cost and is favored due to its minimal emission of pollutants. However, current status reveal its high capital and fuel costs. Nonetheless, mass production is expected to bring the price down and be competitive with the electric vehicle by 2030, as shown in Fig. 4.14b. In terms of performance, FCEVs can compete with ICE in terms of the driving range and the refueling speed. Sustaining the production of FCEVs implicates the necessary establishment of hydrogen filling stations. In this aspect, both ICE vehicles and BEV are more conventional and, obviously, have advantages in terms of refuelling infrastructures. Developments and construction of hydrogen refueling stations, however, are increasing throughout these years. At present, the infrastructures for hydrogen filling stations that are already operating and are planned to be constructed are shown in Fig. 4.15. Most of these infrastructures are situated in

Fig. 4.15 The hydrogen filling stations that are currently in operation and that are planned to be established [59]. Reprinted with permission from [59]. Copyright 2020 by H2stations.org by LBST

developed countries. With the demonstrated advantages of FCEVs, we expect that there will be a worldwide increase in the number of refueling infrastructures in the near future.

4.2 Hydrogen Roadmaps

The abovementioned progress in hydrogen-related technologies stimulated the ongoing attempts to utilize hydrogen as an energy carrier. Several countries are motivated to pursue hydrogen-based economy mainly to solve their dependence on conventional fossil fuels and to alleviate climate change issues due to the emissions from fossil fuel combustion. In this section, hydrogen roadmaps designed to realize hydrogen economy in the near future in different regions are discussed.

4.2.1 Americas

In the Americas, the United States (US) has the most mature infrastructure geared toward a hydrogen economy (for example, see Fig. 4.16). The grand idea of the hydrogen economy development in the US is called the "sector coupling". Sector coupling refers to the concept of interconnecting or integrating the energy-consuming

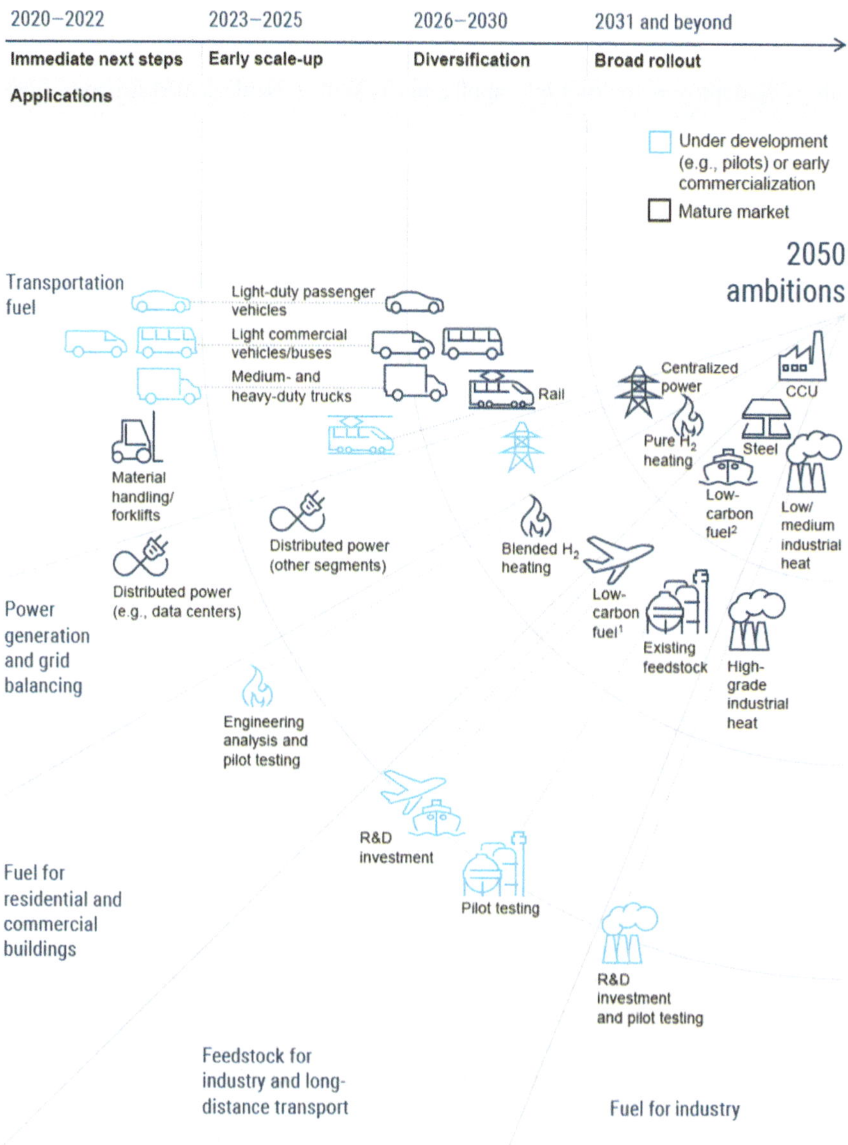

Fig. 4.16 Hydrogen economy roadmap of the US, with the light blue figures indicating technologies that are in there developing/early commercialization stage. The dark blue figures indicate technologies that are already realized [60]. Reprinted with permission from [60]. Copyright 2020 by Fuel Cell & Hydrogen Energy Association

sections—building (heating and cooling), transport, and industry—with the power-producing sector. It is expected that different sectors will benefit when the technology is deployed for multiple applications. With this idea, infrastructure costs could be shouldered by different sectors that will benefit from these technological developments.

As shown in Fig. 4.16, the hydrogen economy roadmap in the US is separated into four key stages: 2020–2022, 2023–2025, 2026–2030, and post-2030. All stages contribute in solidifying a hydrogen economy by 2050. Each stage is accompanied or driven by relevant milestones as shown in Fig. 4.17. During the first stage (2020–2022), it is aimed that hydrogen economy goals be established in more states and at the federal level. The US government will provide more incentives to attract the public sector to adopt the hydrogen technology. Specifically, it is aimed to provide mature hydrogen solutions to market, increase public awareness and acceptance, and continue to pilot the hydrogen use in other applications. The government currently has the niche market for the FCEV forklifts, in which the policy is to encourage the expansion of the FCEV utilization nationwide. By the end of 2022, it will increase the use of hydrogen across all sectors by one million metric tons as compared to 2019. It is estimated that 30,000 FCEVs could be sold by that period.

The second stage (2023–2025) is the early scale-up stage. In this stage, the US government aims to develop large-scale hydrogen production, which will reduce the cost of the gas down and will kick-off the scaling-up of applications. Large-scale hydrogen production facilities will be built using water electrolysis from renewable sources, gas reforming with renewable natural gas (RNG), and carbon capture and storage (CCS) technologies. Hydrogen, in small percentage, will be blended into gas distribution grids to start the scaling-up of production for building heating purposes. States that will adopt this technology will utilize hydrogen-based transportation, and at the same time, increase the hydrogen fueling infrastructures. It is estimated that 150,000 FCEVs will be sold by the end of 2025.

The third stage (2026–2030) is focused in the expansion of hydrogen utilization in the different sectors and states in the US. In this stage, the government is set to scale-up the hydrogen utilization across the country. The hydrogen production by electrolysis would continue to increase, including the building of hydrogen transmission pipelines to further reduce the cost. With this, hydrogen demand will increase by six million metric tons as compared to 2019. It is estimated that 1.2 million FCEVs will be sold and 4,300 fueling stations are already in operation.

At the last stage (after 2030), a complete deployment of hydrogen as energy carrier should have been realized with competitive cost relative to fossil fuel. The market support given to public could then be lifted. The fossil fuel-based hydrogen production facilities will then be replaced with CCS technologies. The sector coupling would benefit from the hydrogen deployment and could create further synergies that would further lower the cost. The established backbone infrastructures will be composed of low-carbon hydrogen production facilities, hydrogen distribution pipeline network, and a vast fueling station infrastructure network. A wide variety of FCEV models will already be available to meet varying customers' needs.

	Today	2022	2025	2030
	Immediate next steps	Early scale-up	Diversification	Broad rollout
H_2 demand, metric tons	11 m	12 m	13 m	17 m
FCEV sales	2,500	30,000	150,000	1,200,000
Material-handling FCEVs	25,000	50,000	125,000	300,000
Fueling stations[1]	63	165[2]	1,000[2]	4,300[3]
Material-handling fueling stations[4]	120	300	600	1,500
Annual investment		$1 bn	$2 bn	$8 bn
New jobs[5]		+50,000	+100,000	+500,000

1 Includes both fueling stations in operation and in development
2 Stations of 500 kg/day; does not include material-handling fueling stations
3 Stations of 1,000 kg/day; does not include material-handling fueling stations
4 Data from Plug Power
5 Includes direct, indirect, and resulting jobs, building on an estimated 200,000 jobs in the sector today

Fig. 4.17 Hydrogen economy milestones of the US [60]. Reprinted with permission from [60]. Copyright 2020 by Fuel Cell & Hydrogen Energy Association

4.2.2 Asia

Asia is the world's largest continent (30% of the Earth's land area) and has the largest population (60% of the total population) [61]. Also, it is regarded as a fast-growing economy region and is projected to dominate the global economy in 2050. This current economic trend will result to a shift or balance of the world's economic power [62]. To sustain its status and satisfy the needs of its people, it is necessary to establish sustainable energy sources. While it is considered as the world's leader on renewable energy, Asia strongly relies on conventional fossil fuels for transportation, industrial, cooling and heating purposes. For instance, renewable energy shares in the total final energy consumption (TFEC) of China and India in 2016 are less than 20 and 40%, respectively [63]. Given this background and the depleting supply of conventional fossil fuels, several countries in Asia are now developing hydrogen-based economy to address such issues.

Japan is among the countries that globally leads in realizing hydrogen-based economy. This is driven by its very limited supply of conventional fossil fuels, in which, it imports ~90% of its primary energy supply. The shutdown of its nuclear power plants since the Fukushima nuclear disaster in 2011 further boosted its import of oil-based fuels. In addition, Japan's advocacy to fight climate change is among the obvious reasons for its desire to shift to a hydrogen-based economy. The country's culture, traditions, and respect for nature is evident from its hosting of the 1997 Kyoto Protocol [64]. The Kyoto Protocol refers to the international agreement which requires the participating nations to reduce greenhouse gas emissions. It put bigger responsibilities on developed nations due to their activities that have a detrimental impact on our atmosphere.

Japan acknowledges the importance of hydrogen as an energy carrier and of reducing carbon emission which leads to the establishment of strategies that will realize the hydrogen economy soon. It aims to (1) diversify the renewable hydrogen source; (2) reduce carbon utilization in transportation, power generation and in industries; (3) realize a society based on hydrogen to achieve energy security, economic efficiency, and environmental protection without compromising safety (3E+S); (4) contribute globally through innovation; (5) expand the available technology domestically and outside the country; and (6) lead the world in realizing hydrogen-based economy. In order to address these goals, several strategies were developed. These include the (1) realization of low-cost hydrogen supply and utilization; (2) expansion of renewable energy for hydrogen production; (3) development of safe and carbon-free hydrogen supply chains; (4) utilization of hydrogen in power generation and in transportation (increase the construction of hydrogen refueling stations and the production of hydrogen vehicles).

Based on its strategic roadmap (Fig. 4.18), Japan desires to reduce the prices of fuel cell vehicles and to lower the operating, construction and component costs of refueling stations. It also targets to reduce the cost of hydrogen production, of carbon capture storage, of water electrolyzer systems, and of fuel cell systems; to

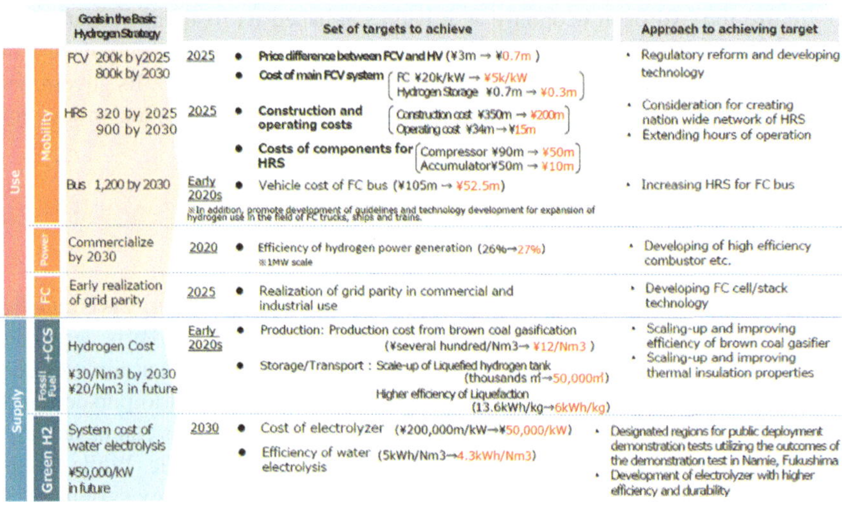

Fig. 4.18 Japans's strategic hydrogen roadmap [65]. Reprinted with permission from [65]. Copyright 2020 by Ministry of Economy, Trade and Industry of Japan

increase the efficiency of power generation, and to scale-up its hydrogen transportation. Such actions likewise acknowledge the difficulties of realizing these goals due to the limitations and costs of available technologies and materials. Nonetheless, it is believed that extensive investigations and partnerships between the government, academe and industries will overcome these challenges. Furthermore, discussion with involved agencies and other countries and acceptance of the hydrogen economy by the society is recognized as crucial factors in realizing these goals.

Japan is currently developing the hydrogen-powered town in Harumi, Tokyo which will accommodate athletes that will participate in the 2020 Tokyo Summer Olympics and Paralympic Games [66]. The buildings/units where athletes will stay will be powered by hydrogen and fuel cell buses will transport the athletes. If this aim of Japan will be successful, it can serve as a benchmark for other nations in its goal to develop a society based on a hydrogen economy.

China is also an active player in the development of a hydrogen-based economy. The country aims to set-up 1000 refueling stations by 2030 [67]. In particular, large investments in hydrogen sector implicates the desire of its government to increase the share of hydrogen in their energy needs. There is increased activity on the establishment of a hydrogen society with its increase in the production and construction of hydrogen buses, trucks, and fuel cells. In fact, the significance of hydrogen in China's energy sector is acknowledged in its future direction ("Made in China 2025" and "13th Five-Year Plan"). Aside from increasing the number of refueling stations, the country desires to increase the number of hydrogen vehicles soon. China's government support to hydrogen-related technology is evident from its subsidy for purchases of hydrogen vehicles, and from incentives and policies with impacts on the realization of a hydrogen-based society [68]. South Korea, on the other hand, also aims to vitalize the hydrogen economy as it had established its roadmap recently [69]. The country focuses on the expansion of its fuel cells and vehicle outputs and the establishment of a system for hydrogen production and distribution. Its government desires the realization of more economic and stable hydrogen supply and production.

4.2.3 Europe

The use of hydrogen at large scale in Europe as a major source of energy is compelled by the recent signing of the COP21 (Conference of the Parties) Paris agreement by 28 Member States of the European Union (EU). The agreement aims to keep global warming "well below 2 °C above preindustrial levels, and to pursue efforts to limit the temperature increase even further to 1.5 °C." [70] This can be achieved by requiring virtually carbon-free power generation at a large scale. While the use of hydrogen is not the only lever in achieving a decarbonized energy system, it has many advantages over the other technologies. It provides a means for large-scale integration of renewables possible because it enables the conversion and storage of energy as renewable gas. It also offers a way to decarbonize segments in power, transport, buildings, and industry, which would rather be difficult to decarbonize.

Generally, the choice of hydrogen as an alternative source of energy in Europe is based on three fundamental arguments: First, hydrogen is the best (or only) choice for at-scale decarbonization of selected segments in transport, industry, and buildings. Specifically, hydrogen is required in the decarbonization of the gas grid that connects Europe's industry and delivers more than 40% of heating in EU. The other levers of decarbonization like biogas utilization and electrification are challenged by their availability at the required scale, cost, and seasonal demands that would require a power storage mechanism at a large scale. On the other hand, hydrogen does not suffer from these shortcomings and can act as a complement to existing heat pumps. Hydrogen can be blended into the existing grid without the need for major upgrades, which can ultimately be converted on pure hydrogen-powered grids. In the area of transport, hydrogen-powered fuel cell is the most promising decarbonization option over batteries and combustion-based energy sources that suffer from lower energy density, high initial costs, and slow recharging performance. Because the transport segment makes up about one-third of all CO_2 emissions in the EU, the use of fuel cells in large vehicles such as trucks, buses, ships, trains, large cars, and commercial vehicles, represents a key element in achieving the energy transition to carbon-free power generation. In aviation, hydrogen and synthetic fuels based on hydrogen are the only at-scale option for direct decarbonization. In the industry sector, hydrogen can be burned to produce high-grade heat or can be used in several processes as feedstock, either directly or together with CO_2 as synfuel/electrofuel, providing a carbon sink, i.e., an opportunity for CO_2 to be used instead of emitted.

The second argument is the systemic role that hydrogen will play in the transition to renewable energy sources, where it is expected to provide a mechanism to flexibly transfer energy across sectors, time, and place. Hydrogen is the only at-scale technology for "sector coupling", which allows the conversion of generated power into a usable form to meet the demand of the end user sectors. It is also the only at-scale technology available for long-term energy storage, addressing the need for increased supply/demand balancing across the year. The use of hydrogen also promises a link between regions, enabling long-distance transportation of energy in pipelines, ships, or trucks, whether gaseous, liquified, or stored in other forms, which costs much less than power transmission lines.

The third argument is that the transition to hydrogen is aligned with customer preferences and convenience, which is important to mitigate customer resistance and adoption difficulties. In transport, for example, hydrogen offers the same range and refueling speed as combustion-engine vehicles. Moreover, energy companies can blend hydrogen or synthetic methane into the gas grip via power-to-gas plants using current piping. While the ultimate goal of 100% switch to hydrogen requires the upgrade of appliances and piping, it still leaves the current heating infrastructure within buildings intact.

The transition to the use of hydrogen as the primary source of energy will radically transform how the EU generates, distributes, stores, and consumes energy. An ambitious scenario for hydrogen deployment in EU is laid in a roadmap based on the inputs from 17 companies and organizations [71]. As shown in Fig. 4.19, the use of hydrogen has the potential of generating approximately 2,250 terrawatt hours (TWh) of energy in Europe by 2050. This represents roughly a quarter of the Europe's total energy demand. This amount of energy is expected to fuel about 42 million large cars, 1.7 million trucks, approximately a quarter of a million buses, and more than 5,500 trains. It is also expected to provide heating to more than the equivalent of 52 million households and provide as much as 10% of building power demand [71]. In the industry sectors, approximately 160 TWh of hydrogen energy would produce high-grade heat, and another 140 TWh would replace coal in steel making processes.

The large-scale deployment of hydrogen in Europe is also expected to stimulate the economy in the region. It would create an estimated EUR 130 billion industry for the fuel and associated equipment for EU companies by 2030, reaching EUR 820 billion by 2050. Such scenario will create a local market for EU industry to use as a springboard for competing globally in the new hydrogen economy. The export potential of EU in 2030 is expected to reach an estimated EUR 70 billion, with net exports of EUR 50 billion. The EU hydrogen industry could provide employment for about 1.0 million highly skilled workers by 2030, reaching 5.4 million by 2050 [71].

In a report prepared for the Fuel Cells and Hydrogen 2 Joint Undertaking, concrete milestones were proposed in the areas of transport, buildings, industry, and in power system. In the transport sector, vehicles powered by fuel cells could account for 1 in 22 passenger vehicles and 1 in 12 of light commercial vehicles sold by the year 2030, leading to a fleet of 3.7 million fuel cell passenger vehicles and 500,000 fuel cell light commercial vehicles. By 2030, about 45,000 fuel cell truck and buses could be on the road, in addition to fuel cell trains that could replace roughly 570 diesel trains. For buildings, hydrogen could replace an estimated 7% of natural gas (by volume)

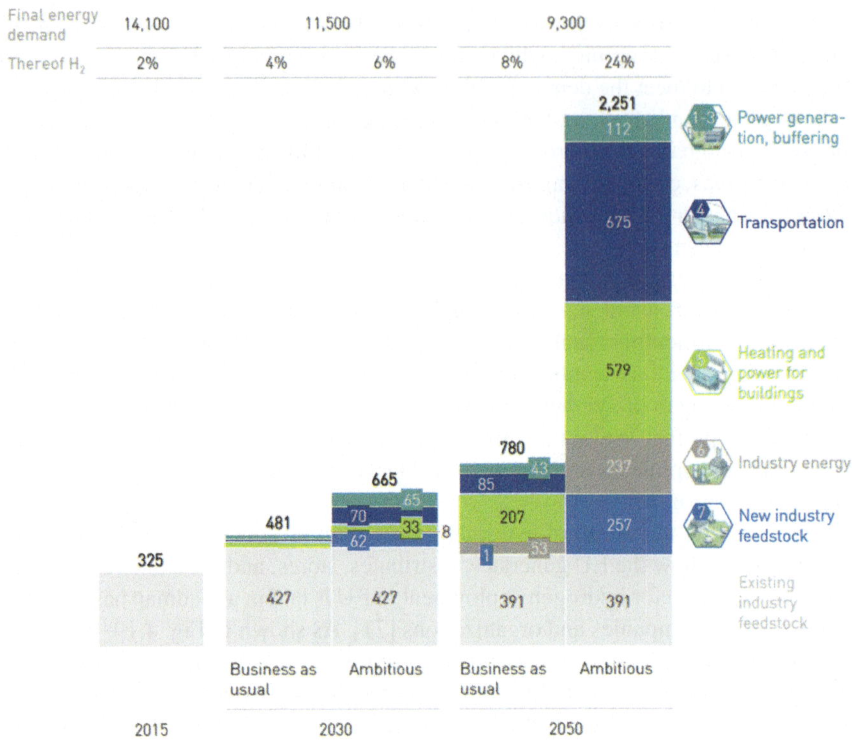

Fig. 4.19 The projected energy generation from hydrogen up to the year 2050. The numbers indicate the energies in terawatt hours (TWh). The "Final Energy Demand" shows the projected energy demand in EU, while the corresponding percentage contributed by hydrogen is shown by the "Thereof H_2". The figure also shows the comparison of the "business-as-usual" scenario and ambitious milestones in 2030 and 2050. Reprinted from [71]. Reprinted with permission from [71]. Copyright 2020 by Fuel Cells and Hydrogen Joint Undertaking

by 2030, and 32% by 2040. It would cover the heating demand of about 2.5 million and more than 11.0 million households in 2030 and 2040, respectively, in addition to commercial buildings. In the industry sector, a transition to one-third ultra-low-carbon hydrogen production could be achieved in all applications, including refineries and ammonia production. In the power system, it is expected that by 2030, the at-scale conversion of "surplus" renewables into hydrogen, large-scale demonstrations of power generation from hydrogen, and renewable-hydrogen generation plants could take place.

These ambitious milestones can only be realized through the coordinated efforts by policymakers, industry, and investors. A much lower level of hydrogen deployment will be achieved if a high level of cooperation does not emerge and current policies remain in place. Such "business-as-usual" scenario is compared to the ambitious milestones in Fig. 4.19.

Different countries worlwide are taking the hydrogen economy seriously. We can see from these roadmaps that such goal will become a reality soon, as driven by the need to replace conventional fossil fuels. While there are current activities in the developed countries to realize such goals, as a result of government, academes, and industry partnership, it is important that public acceptance should be taken into consideration. For example, the use of hydrogen at present is only popular in refineries and fertilizers related applications, and hydrogen as an energy carrier remains an improbable goal for developing countries. Public knowledge on the benefits and advantages of hydrogen in energy-related applications could increase the demand which could boost activities and attract hydrogen investors. The activities currently taking place in developed countries will serve as models for other countries. This could motivate developing countries to consider establishing a hydrogen economy in the future. Or, it could drive developing countries to make initiatives on contributing to the realization of these goals, either in the development of hydrogen infratractures or in providing sustainable hydrogen source. A successful demonstration (cost-effective) of prototype societies / small cities based on the hydrogen economy will certainly motivate other nations to craft their own hydrogen roadmaps.

References

1. J. Rigden, *Hydrogen: The Essential Element* (President and Fellows of Harvard College, USA, 2002)
2. O.T. Benfey, J. Chem. Educ. **29**, 78 (1952)
3. U.S. Department of Energy's Office of Energy Efficiency and Renewable Energy. s. l: https://www.energy.gov/eere/fuelcells/fuel-cell-technologies-office, Accessed October 2019
4. C. Zou, Q. Zhao, G. Zhang, B. Xiong, Natural Gas Indus B **3**, 1–11 (2016)
5. International Energy Outlook 2019 with projections to 2050. https://www.eia.gov/outlooks/ieo/pdf/ieo2019.pdf. Accessed October 2019
6. J.O.M. Bockris, Int. J. Hydrogen Energy **38**, 2579–2588 (2013)
7. G. Zini, P. Tartarini, Solar hydrogen energy systems: science and technology for the hydrogen economy. Springer Science & Business Media (2012)
8. M.H.Rashid, Electric Renewable Energy Systems. Academic Press (2015)
9. B. Ruscic, A.F. Wagner, L.B. Harding, R.L. Asher, D. Feller, D.A. Dixon, K.A. Peterson, Y. Song, X. Qian, C.Y. Ng, J. Liu, J. Phys. Chem. A **106**(11), 2727–2747 (2002)
10. A. Le Gal, S. Abanades, J. Phys. Chem. C **116**, 13516–13523 (2012)
11. R.B. Diver, J.E. Miller, M.D. Allendorf, N.P. Siegel, R.E. Hogan, J. Sol.Energy Eng. **130**, 041001 (2008)
12. C.N.R. Rao, S. Dey, Proc. Natl. Acad. Sci. **114**, 13385–13393 (2017)
13. T. Jafari, E. Moharreri, A.S. Amin, R. Miao, W. Song, S.L. Suib, Molecules **21**, 900 (2016)
14. Y. Asada, J. Miyake, J. Biosci. Bioeng. **88**, 1–6 (1999)
15. S. Ye, R. Wang, M.Z. Wu, Y.P. Yuan, Appl. Surf. Sci. **358**, 15–27 (2015)
16. K. Maeda, K. Domen, J. Phys. Chem. Lett. **1**, 2655–2661 (2010)
17. F.M. Sapountzi, J.M. Gracia, H.O. Fredriksson, J.H. Niemantsverdriet, Prog. Energy Combust. Sci. **58**, 1–35 (2017)
18. A. Li, H. Ooka, N. Bonnet, T. Hayashi, Y. Sun, Q. Jiang, C. Li, H. Han, R. Nakamura, R. Angew. Chem. **131**, 5108–5112 (2019)
19. B. Hinnemann, P.G. Moses, J. Bonde, K.P. Jørgensen, J.H. Nielsen, S. Horch, I. Chorkendorff, J.K. Nørskov, J. Am. Chem. Soc. **127**, 5308 (2005)

20. T.F. Jaramillo, K.P. Jørgensen, J. Bonde, J.H. Nielsen, S. Horch, I. Chorkendorff, Science **317**, 100–102 (2007)
21. Y. Yan, B.Y. Xia, B. Zhao, X. Wang, J. Mater. Chem. A **4**, 17587–17603 (2016)
22. L.A. King, M.A. Hubert, C. Capuano, J. Manco, N. Danilovic, E. Valle, T.R. Hellstern, K. Ayers, T.F. Jaramillo, Nat. Nanotechnol. **14**, 1071–1074 (2019)
23. L. Wei, S. Xu, L. Zhang, C. Liu, H. Zhu, S. Liu, Int. J. Hydrogen Energy **32**, 24–31 (2007)
24. B. Moghtaderi, Fuel **86**, 2422–2430 (2007)
25. S. Farzad, M.A. Mandegari, J.F. Görgens, Biofuel Res. J. **3**, 483–495 (2016)
26. N.A. Al-Mufachi, N.V. Rees, R. Steinberger-Wilkens, Renew. Sustain. Energy Rev. **47**, 540–551 (2015)
27. W. Dong, J. Hafner, Phys. Rev. B **56**, 15396–15403 (1997)
28. S. Sakong, C. Mosh, A. Lozano, H.F. Busnengo, A. Groß, Lowering energy barriers in surface reactions through concerted reaction mechanisms. ChemPhysChem **13**, 3467–3471 (2012)
29. A.A.B. Padama, B. Chantaramolee, H. Nakanishi, H. Kasai, Hydrogen atom absorption in hydrogen-covered Pd(110) (1X2) missing-row surface. Int. J. Hydrogen Energy **39**, 6598–6603 (2014)
30. A.A.B. Padama, H. Kasai, J. Alloy. Compd. **645**, S123–S127 (2015)
31. R.J. Behm, V. Penka, M.G. Cattania, K. Christmann, G. Ertl, J. Chem. Phys. **78**, 7486–7490 (1983)
32. B.D. Adams, A. Chen, Mater. Today **14**, 282–289 (2011)
33. L. Semidey-Flecha, D.S. Sholl, J. Chem. Phys. **128**, 144701 (2008)
34. S. Nayebossadri, J.D. Speight, D. Book, ACS Appl. Mater. Interfaces. **9**, 2650–2661 (2017)
35. Y.W. Budhi, H. Rionaldo, A.A.B. Padama, H. Kasai, I. Noezar, Int. J. Hydrogen Energy **40**, 10081–10089 (2015)
36. Y.W. Budhi, I. Noezar, F. Aldiansyah, P.V. Kemala, A.A.B. Padama, H. Kasai, Int. J. Hydrogen Energy **36**, 15372–15381 (2011)
37. A. Basile, F. Gallucci, S. Tosti, Synthesis, Characterization, and Applications of Palladium Membranes, Membrane Science and Technology, vol 13 (Elsevier, 2008), pp 255–323
38. F. Gallucci, E. Fernandez, P. Corengia, M.S. Annaland, Chem. Eng. Sci. **92**, 40–66 (2013)
39. M.D. Dolan, J. Membr. Sci. **362**, 12–28 (2010)
40. S. Kozhakhmetov, N. Sidorov, V. Piven, I. Sipatov, I. Gabis, B. Arinov, J. Alloy. Compd. **645**, S36–S40 (2015)
41. K. Ishikawa, H. Habaguchi, N. Obata, Y. Kobori, N. Ohtsu, K. Aoki, Int. J. Hydrogen Energy **41**, 5269–5275 (2016)
42. N. Ohtsu, K. Ishikawa, Y. Kobori, Appl. Surf. Sci. **360**, 566–571 (2016)
43. T.L. LeValley, A.R. Richard, M. Fan, Int. J. Hydrogen Energy **39**, 16983–17000 (2014)
44. H. Wang, D.W. Blaylock, A.H. Dam, S.E. Liland, K.R. Rout, Y.A. Zhu, W.H. Green, A. Holmen, D. Chen, Catal. Sci. Technol. **7**, 1713–1725 (2017)
45. Z.X. Yu, D. Chen, B. Totdal, A. Holmen, J. Phys. Chem. B **109**, 6096–6102 (2005)
46. R.L. Arevalo, S.M. Aspera, M.C.S. Escaño, H. Nakanishi, H. Kasai, Scientific Reports **7**, 13963 (2017)
47. I. Dincer, C. Acar, Int. J. Hydrogen Energy **40**, 11094–11111 (2015)
48. M. Ball, M. Weeda, Int. J. Hydrogen Energy **40**, 7903–7919 (2015)
49. R. Moradi, K.M. Groth, Int. J. Hydrogen Energy **44**, 12254 (2019)
50. C. White, R. Steeper, A. Lutz, Int. J. Hydrogen Energy **31**, 1292 (2006)
51. R. Ahluwalia, T. Hua, J.-K. Peng, S. Lasher, K. McKennery, J. Sinha, M. Gardiner, Int. J. Hydrogen Energy **35**, 4171 (2010)
52. Board CAR. Staff report: Initial statement of reasons. (Online). https://www.arb.ca.gov/regact/2011/soreci2011/soreisor.pdf.
53. I. Staffell, D. Scamman, A.V. Abad, P. Balcombe, P.E. Dodds, P. Ekins, N. Shah, K.R. Ward, Energy Environ. Sci. **12**, 463–491 (2019)
54. S.Y. Wang, S.P. Jiang, Natl. Sci. Rev. **4**, 163–166 (2017)
55. A. Tressaud, Fluorine: a paradoxical element, vol 5 (Academic Press, 2018)

56. J.K. Nørskov, T. Bligaard, B. Hvolbæk, F.A. Pedersen, I. Chorkendorff, C.H. Christensen, Chem. Soc. Rev. **37**, 2163–2171 (2008)
57. B.G. Pollet, I. Staffell, J.L. Shang, Electrochim. Acta **84**, 235–249 (2012)
58. P.E. Dodds, P. Ekins, Int. J. Hydrogen Energy **39**, 13941–13953 (2014)
59. Hydrogen Stations Maps (Online). https://www.h2stations.org/stations-map/
60. Road Map to a US Hydrogen Economy (Online) https://www.fchea.org/us-hydrogen-study
61. Asia: Resources (Online). https://www.nationalgeographic.org/encyclopedia/asia-resources/
62. How Asia became a global economic powerhouse (Online). https://www.asiatimes.com/2019/10/article/how-asia-became-a-global-economic-powerhouse/
63. Asia Renewable Energy Report (Online). https://www.ren21.net/asia-report-2019/
64. United Nations Climate Change: What is the Kyoto Protocol? (Online). https://unfccc.int/kyoto_protocol
65. METI. Formulation of a New Strategic Roadmap for Hydrogen and Fuel Cells (Online). https://www.meti.go.jp/english/press/2019/0312_002.html.
66. Tokyo Olympics shine light on hydrogen (Online). https://www.shell.com/inside-energy/japan-tokyo-olympics-hydrogen.html
67. China Hydrogen and Fuel Cell Strategy. (Online). https://www.climate-change-solutions.co.uk/wp-content/uploads/2019/03/WS1AMSpeaker4RalphClagueJLR.pdf
68. Hydrogen Energy Country (Online). https://www.energy.gov/sites/prod/files/2018/10/f56/fcto-infrastructure-workshop-2018-4-li.pdf
69. Korean gov't to promote hydrogen economy through car, fuel cell sectors (Online). https://english1.president.go.kr/Media/News/519.
70. Paris Agreement, United Nations (2015)
71. Hydrogen Roadmap Europe, Fuel Cells and Hydrogen 2 Joint Undertaking, (2019)

Index

© Springer Nature Singapore Pte Ltd. 2020

H. Kasai et al., *Hydrogen and Hydrogen-Containing Molecules on Metal Surfaces*, Springer Series in Surface Sciences 71,
https://doi.org/10.1007/978-981-15-6994-4

The manufacturer's authorised representative in the EU is Springer Nature Customer Service Centre GmbH, Europaplatz 3, 69115 Heidelberg, Germany. If you have any concerns regarding our products, please contact ProductSafety@springernature.com

Printed and bound by CPI Group (UK) Ltd, Croydon, CR0 4YY
30/09/2026
02225515-0009